4/25/22

Quest for Flight

For
Liezel
Compliments of the author

Quest for Flight

*John J. Montgomery
and the Dawn of Aviation
in the West*

BY CRAIG S. HARWOOD
AND GARY B. FOGEL

University of Oklahoma Press : Norman

ALSO BY GARY B. FOGEL

Wind and Wings: The History of Soaring in San Diego (San Diego, Calif., 2000)

LIBRARY OF CONGRESS CATALOGING-IN-PUBLICATION DATA

Harwood, Craig S., 1960–
    Quest for flight : John J. Montgomery and the dawn of aviation in the West / by Craig S. Harwood and Gary B. Fogel.
        p. cm.
    Includes bibliographical references and index.
    ISBN 978-0-8061-4264-7 (cloth)   ISBN 978-0-8061-6475-5 (paper)   1. Montgomery, John J. (John Joseph), 1858-1911.   2. Aeronautics—United States—Biography.   3. Air pilots—California—Biography.   4. Inventors—California—Biography.   5. Aeronautics—California—History.   6. Gliders (Aeronautics)—California—History.   7. Aeronautics—United States—History.   I. Fogel, Gary, 1968–   II. Title.
    TL540.M62H37 2012
    629.13092—dc23   2012002919

The paper in this book meets the guidelines for permanence and durability of the Committee on Production Guidelines for Book Longevity of the Council on Library Resources, Inc. ⊗

To Dyane, Avonlea, and Marilla

CSH

To Eva Fogel, for her continued perseverance and dedication

GBF

CONTENTS

ILLUSTRATIONS

PREFACE AND ACKNOWLEDGMENTS

At the close of the nineteenth century, the collective effort of humankind to conquer the skies constituted a compelling human drama and one of the most important developments in the history of science and technology. That history, however, tends to be represented in the literature from the perspective of those who capitalized on scientific invention. Narratives on the history of human flight often begin in the eastern United States with the emergence of the Wright Brothers and Glenn H. Curtiss, and on the West Coast with Glenn L. Martin—the first successful capitalists in American aviation. Following the sensational public flights in 1908–1909 by the Wrights and Curtiss, the American public showered their heroes with awards, embracing a patriotic belief that Americans had conquered the age-old dream of human-controlled flight. Taking advantage of the wartime need for aircraft, in 1916 the Wright Corporation merged with the Glenn L. Martin Aircraft Company and, along with the Glenn H. Curtiss Company, dominated both the business of aviation and the resulting narrative of aviation history.

Over time, the retelling of this history became increasingly simplified. That was particularly true in the case of the Wright Brothers, whose tireless promotion of a narrative of "first to fly" or "the inventors of the airplane" was gradually accepted by society, with the personal stories and accomplishments of many of their predecessors obscured and contextualized in terms of the Wrights' successes. There may be no better example of this bias than the story of John J. Montgomery, California's "Father of the Aeroplane." Montgomery's independent path to solving the problem of aerial navigation predated and paralleled similar efforts by enthusiasts based in the eastern United States.

This book explores Montgomery's role in aviation through an objective accounting of his life in the context of his background and personal circumstances, interaction with peers, successes and failures, and general isolation in California. It examines society's attitude toward what was then referred to as "the impossible art" of aviation and shifts the traditional focus of American aviation history from the East to the West, revealing Californians' many underappreciated contributions to achievement in both lighter-than-air and heavier-than-air flight.

Quest for Flight speaks to the larger subject of the influences of society and the human condition in the development of science and aviation technology in the Gilded Age in America. It is our hope that this book will not only increase awareness of these early aviation pioneers but also help stimulate further interest in and objective research on Montgomery's leadership and all that he and his peers accomplished during this critical period of American aviation history.

Researchers interested in learning about John J. Montgomery's life have few resources to which they can turn. By far the most comprehensive collection of original source material regarding Montgomery exists at Santa Clara University. Other collections exist at the University of North Carolina, Chapel Hill, and at the National Air and Space Museum Library and the Library of Congress in Washington, D.C. The only noteworthy biography of Montgomery was authored by Arthur D. Spearman more than three decades ago. The advent of the Internet has enabled researchers to gain access to a wide variety of bibliographical and genealogical resources from the period of Montgomery's lifetime, which have provided a new perspective on his life, work, and legacy.

The research required for this book was far-reaching and exhaustive, and we therefore wish to thank the many people who helped make it possible. In addition to the numerous archival facilities visited during the course of this project, we would like to applaud the dedicated staff at the numerous public libraries we made use of during our research. We appreciate the generous offer of historic photographs from the San Diego History Center, the Smithsonian National Air and Space Museum, the Glenn H. Curtiss Aviation Museum, Bruce Coons of the Save Our Heritage Organization, the South Coastal Information Center at San Diego State University, and Santa Clara University Archives and Special Collections. We are particularly indebted to Anne McMahon, Deborah Whiteman, and Sheila Conway of the Archives Office, Santa Clara University, for their professionalism and patience in accommodating our many research visits to the John J. Montgomery Collection during the decade required to research the book.

For their support, enthusiasm, and advice during this project, we would like to thank Carroll F. Gray, John H. McMasters, and Kristen Aliotti. (We note with regret the passing of Dr. McMasters on February 13, 2008.) We greatly appreciate the critical review of this manuscript and the insightful commentary offered by aerodynamicist Paul E. Dees of the Boeing Company; sailplane veteran Richard Huppertz; aerodynamicist Alan C. Brown, former Director of Engineering for Lockheed's F-117A stealth fighter program; Peter J. Westwick, Professor of History at the University of California, Santa Barbara; and the President of Santa Clara University, Michael E. Engh, S.J. We are particularly indebted to Santa Clara University Professor of History Robert M. Senkewicz, S.J., for his crucial support of the project. In addition, we would like to thank Jane Lyle for her excellent editorial assistance on the manuscript, as well as Bob Clark and the staff at the University of Oklahoma Press.

Several individuals and organizations are to be thanked for their interest in Montgomery's legacy, including William J. Adams, Jr., Frank Allen, Dr. Mark Ardema, Raul Blacksten, John Bolthouse, Howard "Ace" and Judy Campbell, Bruce Coons and the Save Our Heritage Organization, Dave Cortese, Dr. Gary A. Flandro, C. Joy Haas, David Hall, John Hibble, N. Ron Hurov, Dr. George Jutila, Bill Liscomb, Ted MacKechnie and the staff of the National Weather Service in San Diego, Tom Martin, former Transportation Secretary Norman Mineta, Godfrey Mungal, Bertha Ryan, Dr. Terry E. Shoup, Chris Silva, Jeff Sinsay, North E. West and the staff at the Hiller Aviation Museum, and Kathleen Winters.

Researchers interested in learning more about John J. Montgomery should visit the world-class research facility at Santa Clara University Archives and Special Collections, and their John J. Montgomery Collection. This definitive and unique assemblage of primary and secondary sources is an essential destination for anyone wishing to explore Montgomery's life. With the publication of this book, we have donated additional primary source materials along with previously unpublished photographs to the Archives in an effort to consolidate the data and bring the collection up to the breadth of our decade of research that has culminated in this book.

Lastly, we would like to thank the following family and friends for support: Lawrence and Eva Fogel, David Fogel, Jacquelyn Fogel, Joanne Fogel, Dyane L. Harwood, and Marcia Jane Wendt (Steffey).

Craig S. Harwood, Santa Cruz County, California
Gary B. Fogel, San Diego, California

Quest for Flight

CHAPTER 1

The Allure of California

California gold was a hidden treasure. But California, like a beautiful maiden—perfect in form and feature, adorned with the matchless charm of innocence, and all unconscious of her own beauty—needed not the vain glitter of gold to complete her loveliness.

Zachariah Montgomery, 1882

John J. Montgomery's history begins with the very independent paths that led his parents, Eleanor "Ellen" Bridget (Evoy) Montgomery and Zachariah "Zach" Montgomery, to California.[1] Ellen Evoy was born on April 23, 1828, in the village of Rosegarland, County Wexford, Ireland. She was the fourth child of James and Bridget Miranda Evoy, who were devout Catholics. Although little is known about their life in Ireland, typical residents of the Rosegarland area made their livelihood as farmers. A few weeks after Ellen's birth, the Evoys (James, Bridget, Mary Ann, Margaret, John, and James, Jr.) boarded a ship bound for America, seeking a fresh start. They arrived in Baltimore, Maryland, in mid- to late May. From there they traveled to Hancock Prairie outside of St. Louis, Missouri, and established a farm.[2]

Only two years later, in early August 1830, James Evoy contracted pneumonia and died. Unable to run the farm and raise five children on her own, Bridget moved into St. Louis and leased out the Hancock farm as a source of income. The prospect of supporting a family was difficult, but she persevered. She never remarried. When news of California's gold circulated in the newspapers, she contemplated the promise of another new beginning in California. The allure of the California Territory proved just too tempting, so at the age of fifty-eight, Bridget liquidated all of her real estate assets, purchased two wagons and some animals (horses, oxen, and mules), and outfitted her family with the supplies necessary for the extended trip across the plains. In the late spring of 1849, Bridget, her daughters Ellen and Margaret, her sons John and James, and Margaret's family (John H. Mc-Courtney and the three McCourtney children) headed out on the Emigrant Trail from Independence. (Mary Ann had married a man named Joseph Millikin and chose to stay in St. Louis.) The sleepy, pastoral California Territory, acquired only two years prior from Mexico, was not yet considered a "legitimate" part of the United States.

The vast majority of emigrants were inexperienced and ignorant of the hardships they were likely to face over the hundreds of miles between Missouri and California as they passed through Indian Territory (Nebraska, Montana, Wyoming, Idaho, Utah, and Nevada). Families with children formed a minority of the emigrant population in 1849, and usually took substantially longer than the average 132 days from the Missouri staging points (St. Louis and Independence) to make the trip to California.[3] Slowed by Margaret's young children, the Evoy/McCourtney group took roughly eight months to make the journey. Along the way they endured many difficulties, including perilous river crossings, the loss of cattle to theft or disease and starvation, the potentially hostile disposition of Native Americans en route, and the constant need for potable water. Diaries of fellow travelers and military correspondence reveal that Bridget's group was among the last to cross the northern Sierra Nevada by way of "Lassen's Route" (or Cutoff) in early November that year. As a consequence, they endured the first snowstorms that fell in the Sierras between October 30 and November 2, 1849. They arrived at Sutter's Fort in the Sacramento Valley in mid-November, with Ellen on horseback, in a dust-covered dress and with one shoe missing.[4]

After picking up new supplies at Sutter's Fort, they relocated to the Yuba Settlement. Located at the confluence of the Feather and Yuba Rivers, the settlement was expected to become a strategic trailhead for miners heading for the northern gold fields of the Sierra Nevada. With the little money that remained, they purchased a lot and constructed a primitive dwelling. The rush of hopeful prospectors heading up the Feather River created many business opportunities. Bridget and Ellen ran a general merchandise business and also sold real estate. Ellen's brothers John and James joined in the search for gold, hoping like so many others to "make their pile." John and Margaret McCourtney established a trading post and toll bridge on the lower reaches of the Bear River, just west of Marysville. Over the next several months, the Evoys settled into the rhythm of daily life in this bustling, raucous gateway to the northern gold fields.

A similar path brought John Montgomery's father, Zachariah "Zach"

Montgomery, to California. Zach was born in 1825 near Bardstown, Kentucky. His father, Thomas Francis Montgomery, was a tobacco and sugar beet farmer who had moved to central Kentucky from Charles County, Maryland, in 1812. In 1815, Thomas met and married Clotilda Wathen in Nelson County, Kentucky. The Montgomerys were southerners from the time Zach's great-grandfather Peter Montgomery emigrated from Normandy, France, to Maryland at the dawn of the eighteenth century. Despite his southern pedigree, Thomas did not keep slaves, although paid African Americans worked on his plantation.[5] Several family members of his generation and his father's generation were prominent Catholics and helped pioneer Catholic missions in central Kentucky. Zach's grandfathers on both sides of the family had served in the Revolutionary War. Clotilda Wathen Montgomery's great-uncle and Zach's namesake, Zachariah Riney, ran a one-room school (the Knob Creek School) at Rineyville, Kentucky, and had served as Abraham Lincoln's first schoolteacher.[6] Ironically, Zach later became a consistent critic of President Lincoln's policies. In regard to this family connection with Lincoln, he once said: "I entertain for my humble, and truly Christian Uncle, a far higher veneration, than I could ever cherish toward his far famed pupil."[7]

While in secondary school at Knottsville, Kentucky, in 1843, Zach chanced upon a book owned by his friend Thomas Bidwell, an account of Thomas's brother John's journey in 1841 across the plains to the Mexican province of Alta California.[8] With its wonderful descriptions of California prior to the Gold Rush, the book had an unexpected influence on eighteen-year-old Zach. "It was by reading this journal of young John Bidwell," he later recalled, "that the mind and heart of this narrator first became fascinated with California." When Zach and Thomas subsequently roomed together at St. Mary's College in Lebanon, Kentucky, he added, "Most of our hours of recreation were spent in each other's company, and in our rambles we often spoke of California, the beauty of its scenery, the salubrity of its climate, the field which it presented for enterprise, the probabilities of its becoming a part of the United States, and the possibility of our making it the place of our future home."[9]

Zach transferred to St. Joseph's College in Bardstown, Kentucky, in 1846, where he received a master's degree in law in 1848. Late that year, news of California's gold reached Kentucky, sparking a great deal of interest and speculation about the prospect of exploiting California's riches. In referring to this phenomenon, Gold Rush historian Jo Ann Levy noted:

> No expression characterized the California gold rush more than the words "seeing the elephant." . . . For gold rushers, the elephant symbolized both the high cost of their endeavor—the myriad possibilities for misfortune on the journey or in California—and like the farmer's circus elephant, an exotic sight, an unequaled experience, the adventure of a lifetime.[10]

As part of an invited speech at St. Joseph's College the following spring, Zach acknowledged the temptations presented by the Gold Rush, but he also appealed to the students' sense of purpose in obtaining an education: "You too are miners, come to dig for gold. Not indeed that base and drossy metal which is the miser's God; at whose shrine fools delight to worship; you are come to dig in the gold region of knowledge, to enrich the precious coffers of your minds, with the precious metals of conscience."[11]

Shortly thereafter, Zach was admitted to the Kentucky bar and served in the moot court of Judge William R. Grigsby in Bardstown. His fascination with California continued to grow, however, and in July of 1850 he decided to head to the California Territory to seek his own share of gold. With borrowed money, he purchased the necessary supplies and met up with a group of like-minded young Kentucky men at St. Joseph, Missouri. They departed St. Joseph on July 31, bound for California on the Emigrant Trail.[12]

After three months on the trail, traveling across the plains and deserts, they finally arrived at "Sacramento City." As Zach described it, "the valley around the city was literally covered with tents and wagons and teams of camping parties, resting from their toilsome journey across the plains, while the roads leading to the city from the various mining localities which made Sacramento their trading point, were alive with teamsters and horsemen driving or riding at a seemingly break-neck speed. Everybody appeared to be in a hurry to 'make their pile' and get out of the country."[13] Eager to start prospecting, Zach departed for Ringgold, a small town located a day's horseback ride from Sacramento.[14] He felt confident that within a day or two, he would be flush with gold. It did not take long, however, for reality to settle in: "It was only the rich strikes that made it into the newspapers, and hence the false idea entertained by most of those who came to California in the sanguine expectation of amassing sudden fortunes by gold-digging. One week's experience at the town of Ringgold was amply sufficient to convince the writer that *her* gold had but a feeble ring for him, and that he must try his luck somewhere else and, perhaps at some other business."[15]

Zach gave up mining and returned to the practice of law. He moved to Sacramento to enter into a legal partnership with John McKune. On September 9, 1850, he took part in the celebration of California's official admittance into the Union, as the thirty-first state. After months surrounded by the bustling, restless stream of humanity that flowed endlessly through Sacramento, Zach grew restless and depressed, a circumstance that was punctuated by the Asiatic cholera epidemic that hit the city in the winter of 1850–51. "For some time the plague raged with such merciless fury that almost the only active business done was the burying of the dead," he recalled.[16] He spoke eloquently of the difficult and lonely circumstances of the illness and death that surrounded him in the "sea of strangers": "The funeral processions were indeed few and small; for, alas! those who would have moistened with tears of affection the cold clay of the dear departed, were far, far away, dreaming perhaps of the rich harvests of treasure their cherished ones were reaping upon California's fabulous fields of gold."[17]

Over the next four years (1850–54), Bridget's youngest daughter, Ellen, continued her career as a businesswoman and a landowner, working as a real estate agent for family-owned properties in Yuba City and neighboring Marysville. Bridget relocated to the settlement of Briggsville in Shasta County, where she opened an eatery and boarding house. Paper money was scarce in the mining communities, so customers often paid in weighed quantities of gold dust. By mid-1853, however, Bridget had grown weary of the chaos and hardships of the gold settlements; so at the age of sixty-two, she sold her Shasta County property, and she and Ellen relocated to Rancho Temescal, a portion of the large land grant owned by José Vicente Peralta, located just north of the hamlet of Oakland.[18] With the gold she had accumulated at Briggsville, she bought a 100-acre tract with an existing residence and farm buildings on what was then called Peralta Road (present-day Telegraph Avenue).[19]

After maintaining law practices in Sacramento and later in Shasta City, Zach tried his hand at placer mining along the Pit River in the Klamath-Sierra region in northern California.[20] In the spring of 1854, he finally concluded that the "elephant" would always elude him. He settled down in Yuba City and established a law practice in the more populated city of Marysville, on the other side of the Feather River. Soon thereafter, he met Miss Helen Francis Graham. After a brief courtship, they married in July 1854. In December 1855 they welcomed a son, whom they named Thomas Graham Montgomery, after Zach's father. Many of Zach's clients were farmers or miners with limited resources, but his law practice was busy. He had finally found a sense of stability and community.

On February 4, 1856, Zach was appointed district attorney of Sutter

County as a replacement for J. S. Reardon, who had abandoned his post. In November of that year, he was elected to serve a full two-year term.[21] He quickly earned a reputation as a vigorous and effective prosecuting attorney. It was said that in his cross-examinations he "was always keen, and if he thought a witness was perjuring himself, he was indeed merciless. Guilty men came to fear his cross-questioning."[22] On July 18, Zach suffered a severe blow when Helen died of an unspecified illness just seven months after giving birth to their child.[23] Her mother cared for young Thomas as Zach struggled to come to terms with his situation.

It is not clear how Ellen and Zach met. Presumably either Ellen or Bridget was conducting some business in Yuba City, where they maintained commercial properties, and somehow made his acquaintance. After a brief courtship, they married on April 28, 1857. Ellen embraced one-and-a-half-year-old Thomas.[24] It was not long before they had children of their own. On February 15, 1858, Ellen gave birth to twin sons, John Joseph Montgomery and Zachariah Montgomery, Jr. Over the next eleven years, they had six more children: Mary was born in 1859, Margaret and Ellen Rose (a second set of twins) in 1861, Richard in 1863, James in 1865, and Jane in 1869.[25] The children were extremely close as they grew up. In fact, Richard, Mary, Jane,

Fig. 1. Montgomery family portrait, early 1859. Left to right: Zach, Jr., Ellen, John, Zach, Thomas. (Courtesy Santa Clara University Archives and Special Collections)

and Margaret, all of whom remained unmarried, chose to live together for much of their adult lives.

At the beginning of the twelfth session of the California legislature (1860–61), Zach won a seat in the California State Assembly representing the 15th District (Sutter County). Believing that parents should have rights with regard to their children's education and should determine how tax revenues were to be spent for education, he made impassioned pleas in speeches throughout the region and in his writings. In 1861, he authored Assembly Bill 348 for educational reform.[26] The "Montgomery Bill" marked the beginning of Zach's campaign to challenge and revise the state's compulsory public education system, a cause that he championed for the rest of his life.

Throughout the 1860s, the Civil War dominated state politics. Newspaper accounts and family correspondence from Kentucky detailing the horrors of the "War between the States" weighed heavily on Zach. With his passion ignited by the war as well as the actions of the Lincoln administration, he embarked on speaking tours throughout California in an attempt to sway public opinion in favor of Confederate secession. Zach's deliberations usually focused on the war's impact on the rights of individual states. On the floor of the state assembly and at public speaking engagements, he engaged in frequent heated debates and point-counterpoint exchanges with some of the most gifted speakers in the region. He was characterized by a fellow assemblyman as a "picturesque character" and a "southern firebrand."[27] One fellow editor, having witnessed several of Zach's more emotional speeches, referred to him as a "human calliope." Zach's leadership and oratory skills were recognized widely by his peers, however, and in February 1861 he was asked to deliver the State of the Union Address for the California legislature.

But tragedy was soon to strike the Montgomery family. On December 28, 1861, six-year-old Thomas and three-year-old Zach, Jr., ventured into a grassy field next to the family home. There they ate some wild mushrooms that they found sprouting from the dewy field. Within hours, both children had succumbed to the mushrooms' deadly poison. Fortunately John had not followed his twin into the field and did not share his brothers' sad fate. The sudden loss of Thomas and Zach, Jr., devastated the young family and generated an outpouring of sympathy and support from the local community.

By the end of 1863, the U.S. Congress and more than a dozen states loyal to the Union had enacted "loyalty oath" legislation. The act was officially designed to "exclude traitors and alien enemies from the courts of justice in civil cases." Many believed that given the requirement to swear allegiance to the United States, lawmakers had gone beyond the normal boundaries of the federal government and ignored the traditional separation of federal and state powers.[28] As a strict believer in the Jeffersonian doctrine of states' rights, Zach refused to take the "test oath," and he quit the practice of law. His growing disdain for the Lincoln administration inspired him to undertake lecture tours of northern California, criticizing the actions of the Lincoln administration and also expounding on the school question. He occasionally received fees for his services.

In March 1864, the Montgomerys suffered the loss of a third child, Ellen Rose, who died from an unknown illness. Yet despite their hardships and losses, Zach and Ellen maintained a nurturing and protective environment for their young family, and a strong faith in God. Zach's personality was at times intimidating and oppressive; however, John and his siblings gained a tremendous sense of self-reliance. Zach had more than a casual interest in the education of his children. John and his siblings were home-schooled through the age of five or six, after which they entered Catholic schools within their local parish. After home schooling, John received his initial education at Notre Dame Academy in Oakland, and at the age of eight or nine, he entered the Christian Brothers School in Oakland, which he attended until the spring of 1874. Given their upbringing, Zach and Ellen placed a high value on Catholic doctrine and parochial education. They viewed the family as its own "little government," with the parents providing moral and legislative guidance. This philosophy was an extension of Zach's belief in the rights of individuals and his critical view of the federal government's compulsory education system.

It was during this same period that the celestial realm first captured the interest of young John. One of his earliest memories was of observing the clouds as they passed by, and in particular how they seemed to rest on the Sutter Buttes, a prominent landmark a few miles northwest of Yuba City that dominates the northern Sacramento Valley: "I thought that if I could get there, I could take hold of the clouds and fly; that I could grasp them and they would carry me with them. I think my interest in aerial navigation dates back to that time."[29] In an environment created and nurtured by the combination of Zach's and Ellen's backgrounds and their philosophies on family, young John Montgomery's natural curiosity flourished. He began to absorb the sights and sounds of the physical world around him and to evaluate how and why things functioned as they did. Through the support and encouragement of his parents, John's curiosity took flight.

CHAPTER 2

The Earth and Vaulted Sky

Reeling from the loss of both profession and family, the Montgomerys moved from Yuba City to Oakland in 1864 to live with Ellen's mother at the Evoy ranch. Bridget and her son-in-law Zach shared an assertive personality and a strong sense of individualism. Zach established the *Occidental* newspaper in San Francisco, chiefly with a view to writing on the school question, but also using it to criticize the Lincoln administration.[1]

At the time, San Francisco was controlled by the U.S. Army, which was headquartered at the Presidio. In the late spring of 1864, the military began arresting prominent dissenting citizens without due process, simply because of their outspoken opposition to the administration. Outraged citizens organized a show of "public indignation" to be held August 3 at Hayes Park. Zach was asked to give an address at the protest, in which he used the unfair actions of the military to reinforce a negative portrayal of President Lincoln. Pro-Union newspapers strongly criticized the meeting, and Zach was singled out for special treatment by the editors of the *San Francisco Flag,* who called for him to be arrested and hanged for treason. Mark Twain attended the event, serving as a reporter for the *San Francisco Daily Call.* In a subsequent article he offered a wry impression of Zach's impassioned oratory style: "After the end of the music . . . Zach Montgomery, of Marysville appeared in the stand. . . . He commenced by saying that he would speak from the record, (thereby meaning that he would read his speech from a manuscript) . . . after a little music to soften the lion which Montgomery had aroused (within himself) . . . Tom Robinson was presented."[2] Zach's published editorials and speeches solidified his reputation among his adversaries as a prominent, if not "rabid," secessionist.

As President Lincoln campaigned for reelection in the fall of 1864, Zach's *Occidental* was one of the so-called Copperhead newspapers that voiced opposition to the administration's policies and demanded that the United States settle with the Confederates. When news of Lincoln's assassination

Fig. 2. Montgomery family portrait at Oakland, ca. 1873–74. Front row (left to right): Richard, Ellen, James, Jane, Zach, John. Back row (left to right): Margaret, Mary. (Courtesy National Air and Space Museum, Smithsonian Institution, SI 75-4245)

reached San Francisco early on the morning of April 15, 1865, an angry mob rampaged through San Francisco, intent on destroying the "Copperhead rags." The offices of the *Occidental* and several other papers, including the *San Francisco Newsletter,* operated by Zach's friend Frederick Marriott, were looted and destroyed. According to family history, Zach narrowly escaped the mob as they entered his building. The fallout from Zach's activities created anxiety for the Montgomerys and further reinforced the importance of a highly protective family environment. For the Montgomery children, the threatening world outside was countered by this family barrier and the moral guidance of their faith.

John inherited a number of traits from his father that influenced the course of his life, including his independence, his obsessive nature, and the easy way he moved from one ambitious project to the next. Occasionally he showed a mischievous side as well. His sister Jane recalled a practical joke that John played on his grandmother Bridget by appealing to her keen interest in mining:

When the men were at lunch, John, armed with a little cannon he had constructed out of lead pipe, loaded it with blueing balls (laundry blueing) and shot his cannon along the newly ploughed furrows—when grandmother saw the blue deposit on the furrows she became very much excited—thought there must be some new mineral in the soil and immediately ordered specimens sent to be assayed—Johnny kept MUMB [sic].[3]

John also enjoyed regaling his younger siblings with tales from his childhood. Jane later wrote, "[John said] he was always getting into mischief—mother would report to father and father would punish him. Finally father took Johnnie to the library and said 'Johnnie I have decided not to whip you anymore—I am going to pray for you.' John said that that appealed to him—whipping him made angry—but the idea that of father praying for him touched his heart—so he gave no further trouble from that time on."[4]

By the age of seven or eight, John had developed an interest in kites. He would ask his mother to help with his kite-making, despite her more serious occupations.[5] He also enjoyed playing with a "flying top," a simple wooden toy consisting of a fixed propeller attached to a stick that when spun rapidly through the hands could achieve lift. The seeds were being planted for his future life's work. However, it was an event in July 1869 that truly fueled John's interest in aerial navigation. It was then that he experienced his first brush with an actual flying machine.

In San Francisco, Zach's friend and fellow newspaper editor Frederick Marriott (1804–84) had designed and constructed a controllable, lighter-than-air dirigible, *The Avitor Hermes Jr.* Marriott's interest in flying machines had originated in his native England while he was associated with the English aviation pioneers William Hensen and John Stringfellow. He had conducted initial indoor experiments with a small-scale prototype of the airship (*Avitor No. 1*) in January 1867.[6] On the strength of these tests, he and his assistants constructed a larger version for outdoor experimentation, *The Avitor Hermes Jr.*, with funds from the Aerial Steam Navigation Company of San Francisco. During the first week of July 1869, they conducted unmanned test flights inside the Avitor Works, an enclosed building at Shell Mound Park, located near the Seventeen Mile House stagecoach stop at Millbrae, just south of San Francisco. One hundred guests were invited to this event, including Zach Montgomery. Given John's interest in flight, Zach returned home with a photograph of the airship.[7] John pleaded to see it in person. On the 4th of July, Marriott gave a successful outdoor public demonstration of *The Avitor* (unmanned and tethered) at Shell Mound Park, with both Zach and John in attendance.[8] Eleven-year-old John was fascinated, and upon returning home, he immediately built a small-scale model of *The Avitor* complete with undercarriage and wheels. He tried to lift a hatchet with its buoyancy, but the model did not have sufficient lift.

In 1869, the California legislature repealed the "loyalty oath" legislation. As a result, Zach resumed the practice of law, initially in San Francisco and then in Oakland, where he quickly established himself as a prominent attorney. During that same year, a French expatriate named Peter Wude stayed with the Montgomerys in Oakland. Wude had survived the siege of Paris during the Franco-Prussian War in 1871.[9] The Montgomery children, particularly John, were captivated by his stories of Parisians using hot-air balloons to remove French government officials during the siege. John's interest in the possibilities of manned flight was further buoyed by his interaction with Wude.

During the 1874–75 academic year, John attended Santa Clara College in preparation for enrollment at St. Ignatius College in San Francisco. The student body at Santa Clara was divided into three sections: the senior division, the junior division, and the preparatory division, which contained the youngest students, including sixteen-year-old John. Santa Clara College was known for its strong science curriculum, largely owing to the efforts of the college fathers to meet the needs of a midcentury California culture that was dependent on technology for mining and transportation. Reflecting both the nineteenth century's lively interest in science and the Jesuits' traditional liberal arts curriculum, the department existed chiefly to round out the education of students enrolled in either the classical or the scientific course. Even classicists, it was reasoned, needed a heavy dose of chemistry, physics, and mathematics.[10]

Intrigued with the scientific facilities at Santa Clara, John focused his attention on mathematics and astronomy, receiving awards and medals for various fields of study, as well as honors for his performance in classical studies.[11] He began to experiment with anything that would fly and focused his interest on theories about his observations. He later recalled an important incident that occurred when he was sixteen:

> I was playing in the barnyard, as boys will, shying a piece of tin in the air. Something in the way it came down, curving and apparently resting at different points in the air, arrested my attention. It struck me as weird. Why didn't it come straight down? Why did it curve and turn and settle—seemingly suspended in midair for an instant? . . . It is with this

incident that I connect my first ideas of aerial navigation that were more than merely visionary and esthetic, more than merely the vague desire to fly.[12]

In 1876, John began his undergraduate education at St. Ignatius College.[13] A successful career in science is dependent upon the quality of formal higher education, and the Jesuits were well known for their abilities as educators, especially in the sciences. The priests and brothers, sent from Turin Province in Italy, established excellent reputations for both St. Ignatius and Santa Clara College. St. Ignatius's physics and chemistry laboratories and collections were considered among the finest in America. In 1880, a national survey of science courses offered by 500 American colleges and universities listed St. Ignatius among 120 institutions judged to be "superior." The school's science courses were described as "unusual" and "remarkable."[14] Although Zach wanted his son to study law, John realized fairly quickly that physics held a much greater attraction for him. The science program at St. Ignatius included inorganic and organic chemistry, an exacting course in physics, and mathematics (through calculus). In his first year (1876), John earned academic honors in religious studies, poetry, mathematics, Latin, Greek, and elocution.[15] In 1877, he was awarded the college's silver medal by the Literary and Scientific Department for his excellent scholastic performance.[16]

Fathers Joseph Bayma and Joseph Neri were among the prominent educators at St. Ignatius who helped cement the school's excellent reputation. Bayma was, in the vernacular of the period, a "philosopher and mathematician." In modern terms he would be considered a theoretical physicist. Neri had a formal education in physics and chemistry from his native Italy, and was a pioneer in applications of electricity, especially electric arc lighting. To illustrate his lectures, he used a large electromagnetic device called the Alliance Machine, which had been brought over from France, where it was used for defensive lighting during the second siege of Paris in 1871 in the Franco-Prussian War. The arc light consisted of an electric light regulator for first-class lighthouses, a spherical mirror, and a large Fresnel lens mounted on a rotating table. Neri made several improvements to the device, including strengthening its magnets by supplying electric current from a storage battery. When the device was placed in the college tower, it could project light to the most distant points of San Francisco Bay. In the summer of 1876, Neri and his students (including John Montgomery) provided public exhibitions of the Alliance Machine as part of an annual industrial fair at the California College at Berkeley.[17] Throughout the nineteenth century and well into the twentieth, commencement exercises at St. Ignatius College were a three-day event, including an exhibition of institutional "show and tell," where the Jesuits and students put their work on public display.[18] During the commencement exercises in 1877, John delivered a lecture titled "Bases of Political Economy."

John graduated in 1879 with a bachelor of science degree in physics, receiving honors in both the sciences and the liberal arts.[19] After graduation, he developed a "solar microscope," a device that used sunlight to produce magnified images on walls, with which he entertained friends and family by demonstrating insect anatomy and by showing the vibrations of the human voice through the projection of sound waves created with a tuning fork. He also built a dynamo and began conducting his own experiments with electricity. Zach was impressed with John's ingenious scientific demonstrations. As Jane recalled, "While a student at St. Ignatius College, John made a very impressive looking electrical machine which pleased father. He was amazed and proud of John's skill in cutting the large round glass wheel for his electrical machine [i.e., the dynamo]."[20] Zach invited another friend, Joseph LeConte (1823–1901), to view a demonstration that John gave at the family home in the summer of 1879.[21] LeConte was a prominent educator and geologist and a co-founder (along with his brother John) of the California College at Berkeley, as well as a co-founder of the Sierra Club in 1892. He left with a strongly positive impression of John. Jane recalled, "At one of his exhibitions Professor LeConte . . . predicted a great future for John."[22]

In the fall of 1879, John started graduate studies at St. Ignatius. He was already receiving public acknowledgment for his unusual aptitude in science. A newspaper article from June 1879 noted, "John J. Montgomery, eldest son of Hon. Zach Montgomery, is one of the young men of our city who is bound at no very distant day to make a name for himself as a scientist."[23] His faith in the visionary field of aeronautics set him apart from his peers. His peers, however, were not as quick to recognize his promise. The Reverend Fred Morrison, S.J., a younger student at the time at St. Ignatius, later recalled:

> Among the seniors, or as it was called in those days "The Philosophy Class" was John Montgomery, the famous inventor of the "glider" flying machine. In those days anyone who even mentioned, "man being able to fly" was considered a little bit "off."[24] So, when John was in the vicinity there was a general tapping of heads which in our present day would be the sign that the party was crazy. Nevertheless, he went the even tenor of

his way and proved that we were all mistaken. . . . He was a genius before his time and a credit to the College.[25]

The science program for the master's degree at St. Ignatius included higher mathematics (beyond calculus), Newtonian mechanics, astronomy, biology, analytical chemistry, mineralogy, and geology. John received his master's in June 1880, earning honors for his science and classical studies.[26]

John's primary mentors, Father Bayma and Father Neri, had an indelible influence on him. Neri's pioneering work in electricity, optics, and electromagnetism became a recurring theme in John's life. Bayma's self-reliance and his preference for engaging in visionary research were traits that resonated throughout John's career.[27] During John's attendance at St. Ignatius, both Bayma and Neri told Zach how much they admired John and his natural talent for science and mathematics. Ironically, Bayma also asserted that the impossibility of mechanical (heavier-than-air) flight could be proved mathematically.

In 1881, Zach retired from the practice of law and sold his partnership in Oakland. He established a new monthly serial, *The Family's Defender Magazine and Educational Review,* devoted to educational reform and parental rights in "the school question."[28] After graduating from St. Ignatius, John acted as Zach's personal secretary. He answered correspondence, perused science-themed serials, and flagged potential subject matter for *The Family's Defender.*

Sensing that John needed a career of his own, Ellen decided that he should be a businessman. She went to great lengths to make it happen, including having a wood-frame building constructed on a portion of the Montgomery tract fronting Telegraph Road in Oakland to house a general store. As the manager of "Montgomery Bros.," John tended the register while his younger brother James delivered customers' purchases. However, John's thoughts usually seemed to be elsewhere. At times when he was on duty at the register, he became so engrossed in mathematical calculations that his customers (mainly neighbors who knew him well) simply placed their money on the counter on their way out the door rather than disturb him. His inattention to customers and his genuine lack of interest in the store were a source of amusement among family and friends, but a disappointment to Ellen, the consummate businesswoman.

Ellen suffered from severe asthma while the family lived in Oakland. In need of a vacation, and having read travelogue articles describing the charm and healthful climate of San Diego, she and Zach vacationed there in 1881.

Fig. 3. John J. Montgomery, ca. 1880. (Courtesy Santa Clara University Archives and Special Collections)

They became enamored with the area's climate, idyllic setting, and bountiful agricultural potential. Zach had been thinking about slowing down the pace of his life in a sort of semi-retirement, and the prospect of becoming a gentleman farmer appealed to him. After several inquiries, they discovered that Alfred H. Wilcox, a prominent citizen and the captain of a schooner between San Diego and San Francisco, was interested in selling his homestead on a portion of a large tract known as Rancho La Punta (Melijo) near Otay, a small town near the mouth of Otay Valley. They purchased eighty acres of land from Captain Wilcox and relocated the family to San Diego. Everyone but John moved to Otay in late December 1881.[29]

Otay was a sparsely populated area located between the small, bustling town of San Diego and the Mexican border, a half-day's stagecoach ride from San Diego. On the coastal plain, many acres were developed, with young groves of citrus and experimental plantings of many other varieties of tropical and semitropical fruit trees.[30] Zach anticipated harvesting and selling produce in a self-sustaining operation. Within a matter of months and with the help of hired local Kumeyaay Native Americans, the Montgomerys established fig, olive, guava, grapefruit, peach, orange, and lemon orchards and a vineyard of Muscat grapes. The ranch quickly became known as "Fruitland." Otay Valley was populated largely by conservative Hispanic families, some of whom had lived in the area for generations, engaged in ranching and agriculture. Other families were transplants from the Midwest or the East Coast, lured to the area by the inexpensive real estate and agreeable climate. The Montgomerys befriended their neighbors and became particularly close to the families of William Blount Couts (Thomas, Charles, and Katie), Santiago E. Argüello, David Arnold (Isabel), and David Burroughs (Charles). The Argüellos' home, located just to the north of Fruitland Ranch, was a well-known landmark along the stage line to Baja California.[31] The Montgomerys soon settled into the rhythm of life at Otay, dictated largely by a lifestyle of farming. The only thing missing from their family life was John.

Fig. 4. Fruitland Ranch, Otay Valley, Calif., 1887. This view is toward the north. The white A-frame building in the center enclosed John's blacksmith shop (lower level) and laboratory (upper level). It was here, in the 1880s, that the first successful heavier-than-air flying machines in the Western Hemisphere were conceived of and constructed. The historic Argüello Adobe is in the background. (Courtesy Bruce Coons)

CHAPTER 3

Tutors in the Art of Flying

By early 1882, Ellen had come to the realization that the Montgomery Bros. store was a financial boondoggle. She asked John to close the business in Oakland and join the family in Otay. In February he boarded the schooner *Orizaba* and traveled down the coast to San Diego. Upon arriving at Fruitland, John made improvements to the homestead, including a wood-frame addition to the main residence and a large concrete ornamental pool centered within the turnaround in the front entryway. He assisted Richard for the first few months, platting off the ranch and expanding the vineyard.[1] He hired local Kumeyaay Indians to work under his supervision to maintain and develop portions of the orchards. Once these initial improvements were completed, he established his own laboratory. As Jane recalled, "John's laboratory was in the second story room in a row of buildings about 75 feet from the home. This group of buildings contained the art studio where Mary and Margaret practiced painting in oils and water colors."[2] On the lower level of the building he established a blacksmith shop, including a forge, lathe, and other equipment. As a teenager, John received training in blacksmithing, mechanics, and basic carpentry from his uncle John McCourtney. Prior to coming to California, McCourtney had served as a professional ship's mechanic.[3] These skills would be essential to John in his experimentation with flight and other aspects of science, especially when he was working in isolation.

John was not always hard at work, but it seemed that his mind was always preoccupied. Said Jane:

> When John wanted to quietly study he would go out to the garden and when comfortably seated with his chair tilted back, his legs crossed and his hands clasped behind his head—he would soon be completely lost in study . . . smoking his pipe and thoughtfully and happily watching the smoke rings. . . . John was very fond of music—he knew my practice hour and he would often sit under the arbor directly in front of the house or in the summer house particularly if there were birds flying about. He loved the scales, Czerney's Estudes [*sic*] and Heller's Studie particularly

21

the Allegro movements, also "Tarantelle"—but his favorite piece was "The Visitation Convent Bells" by Kunkle. He said the buzzards seemed to soar in perfect time to that waltz so he said: "Let's change the name to The Buzzard's Waltz."[4]

He occasionally diverted his energies to other areas of interest. With his solar microscope, he examined the wing structure of various flying insects—enlarged many times. He was particularly interested in their structural details and the strength of the wings of the housefly and dragonfly. He also made a gyroscope to prove the correctness of his theory of the tilting of the planets.[5]

There were many demands on John's time during the growing season at Otay. Springtime marked the beginning of a very busy period of planting and irrigating. Maintaining the irrigation systems in the fields was a constant preoccupation throughout the summer. In keeping with the traditional regional pattern for farming, crops were harvested in the fall, followed immediately by tilling prior to winter. Then came the "dull season," when men and their teams of oxen and mules could rest from their labors until the following spring.[6] It was during those months that John was most able to devote time to science, particularly aeronautics, ornithology, and electricity.

Of his interest in ornithology, John said:

> It has always seemed to me that the secret of aerial navigation lay in the discovery of the principle of a bird's flight—that the successful airship would not be modeled on the balloon nor dependent on its buoyancy. The flying of wild geese interested me in consequence and as I couldn't get near enough to them for close and careful observation, I taught my grandmother's flock of geese to fly. I used to drive them down to the extreme end of the ranch and then by cracking a whip behind them compel them to rise and fly to the other end of the ranch. I think it was always a matter of wonderment to my grandmother why her geese could fly so much farther and better than her neighbors'.[7]

From 1882 to 1885 in San Diego, he spent a lot of time studying birds. Watching them fly inspired him to consider whether humans might be able do the same. Hiding behind shrubs or rocky outcrops on the side of Otay Mountain, he observed wild birds up close as they rode the rising air currents nearby.[8]

> The turning or twisting of the wings or tail, the contracting and expanding of the wings and their backward or forward movements, with the

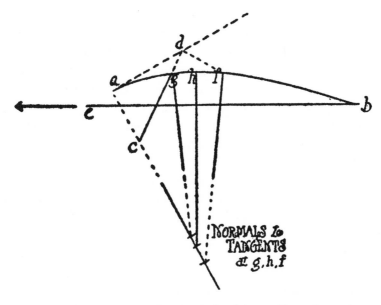

Fig. 5. Diagram by Montgomery showing the breakdown of forces (normal pressures and tangentials at points g, h, and f) and the center of pressure on a cambered wing as determined from ornithological observations (beginning in 1882) and wind tunnel experiments in 1885–93. (From Victor Lougheed, *Vehicles of the Air: A Popular Exposition of Modern Aeronautics with Working Drawings* [Chicago: Reilly and Britton, 1909], 459)

corresponding evolutions, were all easily seen. Sometimes they would engage in play or combat and then their powers of flight, agility and mastery over the air were fully manifest. And their control over every phase of flight, from darting down hundreds of feet below to rising hundreds above, to soar wherever they wished, was so perfect and easy that one irresistibly felt that the mastery over the air must be within easy reach.[9]

He also kept captive birds. After completing duties at the ranch one day in 1882, John and James traveled to the east end of Otay Valley, where they collected an eaglet from a nest on a steep bluff. In addition, according to James, John had "hawks, and a buzzard, heavy birds that he studied constantly in order to formulate his theory upon the question of flying, for the reason that he was of the opinion that if flying were ever to evolve successfully by man it would brought about by following the laws of nature."[10]

For more detailed anatomical studies, John collected specimens of large birds, including eagles, pelicans, and buzzards, that had died of natural causes. Sometimes he hunted his own. After stretching the wings and tacking them on the side of the barn, he let them dry in the sun in an extended, flying position. He performed measurements of the center of gravity by bracing the wings in the normal position for soaring and then balancing the body by thrusting sharp points into it, immediately under the wings. Within a few years, and after testing airfoil shapes with a crude wind tunnel, he applied calculations to determine that this same point coincided closely with the center of pressure on curved surfaces.[11] Additionally, he performed experiments to determine the ratio of wing surface area to lifted weight, and he studied how birds' wings interacted with air currents. In retrospect, John was one of the earliest ornithologists in the region, working independently to gain an understanding of the wing structure and behavior of soaring birds as they might be applied toward a solution for controlled flight by man.

Following a common approach of the day, and in light of his predisposition to mimic what he observed in nature, John's first inclination was to build an ornithopter, a machine with flapping wings. Several Europeans also independently investigated the use of ornithopters during this period, including Jean-Marie Le Bris (1857), Charles Spencer (1868), James Bell Pettigrew (1873), Otto Lilienthal (late 1870s), and an M. Dandrieux (1879) of France.[12] In America, only a handful of experimenters had obtained patents on such devices, including Watson Quinby (1869), François X. Lamboley (1876), and Melville M. Murrell (1877).[13] Even as late as the 1910s, researchers such as Harry La Verne Twining and an experimenter by the name of Quartermaster continued to work with ornithopters. American aviation pioneer Octave Chanute commented in 1894, "Each fresh inventor of winged machines is apt to imagine that his predecessors did not succeed because they did not hit upon the right method of imitating the complicated and swift motions of the birds."[14] These devices were universally unsuccessful. Generally unaware of what other aeronautical workers were doing, John evaluated similar hypotheses in an attempt to raise himself from the ground using muscular effort to drive a flapping set of wings.[15] In early 1883 he built three ornithopters, but none of them were capable of lifting his 130 pounds. Discouraged by the difficulty of flapping flight, he turned his attention to the study of gliding, hoping to solve some of the mysteries of soaring.[16]

James Montgomery later noted that after news of John's ornithopters spread to neighboring ranches, he became the object of considerable derision. Thereafter, his experiments at Otay were conducted in private. Other early aviation pioneers felt the same need to exercise discretion regarding

Fig. 6. John's laboratory at Fruitland, ca. late 1880s. Note the electrical generator ("Dynamo") on the right side of the top shelf, and the gyroscope (marked "X"). (Courtesy Santa Clara University Archives and Special Collections)

their research. By keeping a low profile, they could avoid being ridiculed. Louis-Pierre Mouillard, who made his first successful gliding flight with his *Model No. 3* in 1865 in Mitidja, Algeria, encountered a similar attitude toward his attempts to emulate the birds. On one occasion, his desire to make a flight without becoming a laughingstock at the ranch where he was staying inspired him to employ "a series of profound combinations and pretexts in sending everybody away, so that I was left all alone on the farm. . . . and as soon as their backs were turned, I strolled into the prairie with my apparatus upon my shoulders."[17] After going public with his aeronautical work in 1893, German gliding pioneer Otto Lilienthal commented:

> The great delay in publishing discoveries of myself and [my] brother on aviation was only the natural result of the circumstances under which they were made. Even when we were devoting every hour of our leisure to the question of aviation, and were already on the track of the laws which were to evolve this problem, people in Germany still considered every man who occupied himself with this unprofitable art as little better than a lunatic. This was sufficient cause for our not attracting unnecessary attention to such studies.[18]

Anyone experimenting with flight prior to the turn of the century needed a thick skin, discretion, or both. By the mid-1890s, a more open-minded attitude slowly emerged among experimenters (see chapter 7).

Concurrent with his experiments with ornithopters, John also improved his dynamo, which he had brought with him from Oakland. A particularly powerful prototype that he had developed led him to consider filing a patent. In July 1883, however, Zach showed him an article in *Scientific American* that described a very similar dynamo built by William Thomson and Sebastian de Ferranti wholly independently of John's research.[19] John's frustration was palpable. As his brother Richard recalled, "he was so provoked that he smashed the device with a hammer."[20]

Zach continued to publish *The Family's Defender* from the ranch. His need to handle mass mailings of the magazine, along with a large volume of correspondence, led him to establish the Fruitland post office in March 1882; he maintained it for the next three years, after which it was closed.[21] Occasionally the publication's education-based themes were supplemented with subjects of more casual interest, such as methods of fertilizing soil, pest control in fruit-based agriculture, and advice on irrigation. John continued to serve as Zach's secretary, suggesting additional source material. Thus the journal occasionally included topics of scientific interest, including electrical phenomena, planetary motion, telegraph communications, and patents for both devices and processes. The spring 1882 issue included an anonymously authored article titled "The Ethics of Invention," with particular emphasis on the potential use of flying machines for both nefarious purposes (such as war) and benevolent purposes (such as surveying for mineral deposits and exploring remote lands).[22] It also discussed patent infringement. An earlier article had described a process for the vulcanization of India rubber, methods used in a hot-blast furnace, and the issue of patent infringement, all topics relevant to John's future as an inventor.[23] In fact, while at Fruitland, John devised a novel and effective method for both devulcanizing and vulcanizing India rubber, for which he obtained a patent in 1884, demonstrating his skill as an inventor as well as a clear understanding of the patent process.[24]

John's central passion, the pursuit of aeronautics, was based on his study of birds, his so-called "tutors in the art of flying." Half a world away, Louis Mouillard had similarly referred to the vultures in Algeria as "my little professors." Over a period of months in 1883, John observed the flight patterns of large flocks of pelicans over San Diego Bay.[25] He noted that when they were soaring as a group, their wings were arched downward to the fullest extent, and when viewed from a distance, their column presented a series of waves, with the individuals successively descending and ascending. He shot several pelicans and determined that they weighed an average of 12 pounds and had a wingspan of 9 feet. He concluded, "first, there were twelve hundred pounds sustained and moving along without apparent effort; and, second, this not on a single surface, but several in tandem."[26] He later revealed that it was these particular observations that had inspired him to turn his attention to the concept of fixed-wing flying machines.[27] The tandem-wing concept and downward arch of the wings also were adopted later in his most famous aircraft design. His ornithological investigations served him throughout the remainder of his scientific quest toward human-controlled flight. Although the fertile Otay Valley provided a bounty of agricultural product for its farmers, for John the only real fruit to be cultivated at Fruitland had its roots in scientific invention.

CHAPTER 4

Wings over Otay

A flying machine is impossible, in spite of the testimony of the birds.

Joseph LeConte, 1888

By the time of his move to Otay, Zach had already received accolades for his efforts in law, politics, and writing. His biography was included in many texts on the history of California, including Herbert Howe Bancroft's seminal *History of California* (1885), and he was viewed as an influential and highly individualistic character in California's history. His circumstances would soon have a significant influence on John's course in the then speculative and controversial field of "artificial flight."

After the double disappointment of the ornithopters and his dynamo in midsummer 1883, John recommitted himself to solving the problem of aerial navigation, this time through the use of fixed-wing gliders. His principal aim, according to his brother Richard, was to "solve the problem of balance and equilibrium. He asserted to me time and time again that he could pull a board through the air with power, sufficient to keep it in motion, that of itself it could not maintain equilibrium, and his main thought seemed to be to devise an apparatus that under all circumstances would maintain its balance and equilibrium."[1]

Experimenters in the late nineteenth century commonly failed to recognize the advantages inherent in adopting the cambered or curved shape of birds' wings when designing airfoils. The common belief in the period was that a flat or "planar" airfoil would provide sufficient lift for flight.[2] John, however, had observed that birds' wings were curved and that the curve varied from the base to the tip of the wing, and he trusted that they were arranged in this manner for a reason, although he did not yet understand why in terms of engineering or physics. He later revealed, "At that time I started out on the ideal of most men who have studied aeroplanes and endeavored to imitate a bird's wing, as I perceived it."[3] He added, "With

28

this, I went against my reason, but followed nature."[4] The value of mimicking nature was also noted by Chanute in 1894: "Those who have succeeded in the air, the true experts in gliding, the soaring birds, do not perform their evolutions with plane surfaces. . . . It is true that in many cases they do not differ greatly from planes, and the mind of man so strongly tends to the simplification of complicated shapes, that most inventors have assumed that the effect on the air will be practically the same."[5]

John's first glider was apparently constructed on a part-time basis during the lull on the farm after the 1883 harvest, which extended from the late fall into the spring of 1884.[6] Everyone in the family contributed to the project in some way. Jane, for example, recalled that she and Mary helped John in late 1883:

> I recall John discussing with father the difficulty of curving the wings— and his final decision to steam the wood to make it pliable—he accomplished this by means of a contraption suspended above the blacksmith forge which he equipped with a perforated pipe attached to a water pipe leading down to the bed of the forge where a coil was formed. Water was fed into this coil from a faucet close by. John and I happily blowing the bellows to the tune of "The Anvil Chorus" from "Il Trovatore"—until sufficient steam had been generated to thoroughly limber the sticks. They were then taken out and placed on the walls of his shop, in the desired curve and left to dry. My sister Mary and I under the watchful supervision of John sewed the unbleached cloth to the machine—after which he brushed the cloth with linseed oil—to strengthen as well as to make it as air tight as possible.[7]

Through their assistance, Jane and Mary were inadvertently the first women to participate in the construction of a heavier-than-air flying machine in America. The airfoil used on this first glider was based on John's observations and measurements of the wings of the California gull (*Larus californicus*), a bird that is ubiquitous throughout coastal California. This glider had a wood frame (ash and spruce); the wing was made from oiled cotton fabric, with a wingspan of 20 feet and a chord—the width of the wing from the leading edge to the trailing edge—of 4½ feet. It weighed 40 pounds and had an operable semicircular elevator (tail) connected by a cable to a wooden handle beside the pilot.

Zach had come to the realization that he could do more to bring attention to the cause of "the school question" by reentering politics than by continuing to champion his cause through literature. *The Family's Defender* therefore ceased publication in early 1884, and Zach opened a new law office

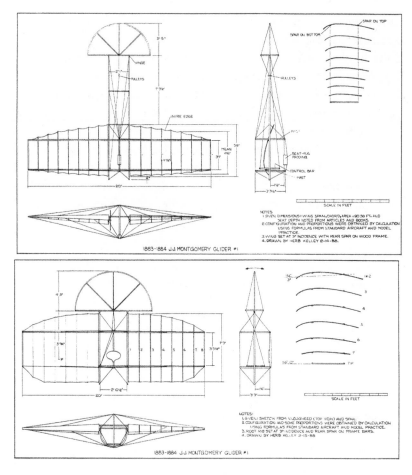

Fig. 7. Herb L. Kelley's study of Montgomery's first fixed-wing glider of 1883–84. The sketch on the bottom is largely a reproduction of Victor Lougheed's rather crude drawing of the craft (1909), with additional assumptions about dimensional details provided by Kelley. Montgomery's original hand-drawn sketch of this machine was misplaced by Lougheed sometime prior to the commencement of a federal court case (Equity No. 33852) filed by the Montgomery heirs in 1919. The top sketch was Kelley's attempt to re-create the craft by extrapolating details from a surviving original hand-drawn figure by Montgomery of his second and third designs of 1885, and 1886, respectively (see figs. 9 and 12). Kelley's drawing is probably more representative of the 1883–84 craft than Lougheed's in terms of fuselage length and minor details. (Courtesy WWI Aero, Inc.)

in San Diego as a secondary source of income to supplement his earnings from the farm. In mid-April he left Fruitland on a six-week lecture tour of California to further promote a reform of the public educational system. After an extended period of record-breaking rainfall from January through mid-March 1884 (a record that still stands in the San Diego region), conditions finally improved.[8] The period after the rain subsided through the time that Zach departed San Diego in mid-April provided John with the first real opportunity in months to try flying his glider. Jane well remembered that "long expected test of the FLYING MACHINE—how John and James slipped away in the early morning with their precious burden perched on a hay wagon."[9] The brothers left before dawn to avoid being seen by their neighbors. They traveled to a spot at the western edge of Otay Mesa that John had selected previously for the flight tests.[10] In a lecture given to the Aeronautic Society of New York in 1910, John recalled:

> I took the machine to the top of a hill, facing a gentle wind and with a little run and a jump I found myself in the air. I proceeded against the wind, gliding downhill for a distance of about 600 yards [*sic*], being able to direct my course at will.[11] A peculiar sensation came over me. The first, in placing myself at the mercy of the wind was that of fear. Immediately after came a feeling of security when I realized the solid support given by the wing-surface and the support was of a very peculiar nature. There was a cushiony softness about it, yet it was firm. When I found that the machine would follow my movements in the seat for balancing, I felt I was self-buoyant.[12]

In several interviews conducted years later, John reported that multiple glides were made with the first glider and that it "would glide 200 yards," with a difference in elevation from the launching and landing points of about 200 feet.[13] Nineteen-year-old James was amused by his older brother's ambitions; in describing his overall impression of the day of experiments he assisted with, he later commented, "It was merely one of the initial experiments to solve the problem. . . . I was a youngster then . . . and [I recall] the ideal [*sic*] of avoiding the neighbors, so they would not see [us], [and I would] have a chance to frolic, and laugh at this would-be aerial navigator."[14] After several flights, however, the experiments came to an abrupt end when John had a hard landing due to a gust of wind, damaging the machine beyond easy repair. They waited on the mesa until dusk, then returned under cover of darkness to the ranch, where their mother and sisters were waiting anxiously. Jane later recalled John's elation at his success.[15]

On this first fixed-wing glider, the pilot controlled roll by shifting his weight, and pitch was controlled through an operable elevator. Control of the yaw axis was left unaddressed. Despite the success, John recognized a flaw with the initial design: "I found that the machine swung side to side in the wind. One or [the] other of the wings would dip, and once the apparatus turned right over. That led me to make a [further] close study of the wings of birds."[16] Several aviation pioneers had already made gliding flights (among them Sir George Cayley's coachman, Jean-Marie Le Bris, and Louis Mouillard), but John's experience was different because of his clear description of controlling the flight.[17] He hypothesized that a solution to the problem of flight would follow from the application of mathematics and physics (concepts he termed "mechanical principles").

Through these initial glides at Otay, John Montgomery was one of only a handful of pioneers to have any success with a man-carrying glider, and in the process he made the first controlled flights of a heavier-than-air flying machine in the Western Hemisphere. Years later, when describing his many experiments in the 1880s, John said that his attempts with flapping wings had been his "only real disappointment."[18] The precedent established with his practical glider experiments was lost on him at the time as he quietly pursued the underlying principles of flight. His first steps on that journey had left him with more questions than answers.

CHAPTER 5

The Internal Work
of the Wind

Neighbors who knew he was working on the flying machine thought he was crazy on the subject, and when he went out to the Otay Mountains to try out his machine he took his rifle to give the impression he was out deer hunting.

Charles Burroughs, 1920

Upon returning from his lecture tour in late May 1884, Zach focused considerable energy on his law practice.[1] With county elections looming, members of the San Diego County Democratic Committee urged him to run for the office of district attorney.[2] Although Zach was well qualified, some local Democrats were wary of his strongly negative stance toward the compulsory public education system. As a result, he ran for office as an Independent. He officially announced his candidacy in the summer of 1884.[3] His opponents went on the offensive, circulating material written by John Swett, the former superintendent of public instruction in California, that accused Zach of introducing a proposition to divert state taxes to Catholic schools.[4] These organized challenges placed his campaign in a precarious state. The growing public interest in Zach and his family came at the same time as John's initial efforts at gliding. Public disclosure of John's "eccentric" activities would undoubtedly have had negative consequences for Zach's political aspirations. John's passion for secrecy thus became even more intense.

Following the end of the academic year in June 1884, Richard returned to Fruitland from Santa Clara College.[5] With the help of some hired hands, he soon established additional orchards and vineyards. Roughly two months later, James departed Fruitland to attend Santa Clara College. He would remain at school (first at Santa Clara College and then at Georgetown College in Washington, D.C.) for the next five years. James was therefore not directly involved with any of John's additional flying experiments beyond that single first day of flights made in the spring of 1884.

Fig. 8. At Fruitland Ranch, ca. 1889. Left to right: James Montgomery, Edwin Belden (a cousin-in-law, seated), Richard Montgomery, John Montgomery. (Courtesy Santa Clara University Archives and Special Collections)

John continued to observe birds at his favorite spot at Otay Mountain in an effort to understand how they coped with the side gusts that had affected his flying machine. He marveled at their ability:

> In the mountain regions, where I performed my first experiments in gliding flight, I have watched for hours at a time the graceful movements of buzzards, in the hopes of catching their secret and imitating them. While I remained hidden in the bushes they would pass within a few feet of me, sometimes just above, then in front and again below me, gliding to and fro, tilting and moving to one side and then the other; sometimes sinking and then rising. Yet all these movements were produced by changes in the adjustment of the wings or body too slight to be detected.[6]

He immediately began designing a second fixed-wing glider to test his theory, but he was unable to build the machine and include the control systems that he envisioned while also using a cambered airfoil. In order to facilitate his design, he resorted to a flat rather than cambered airfoil. He did not yet fully understand the most efficient form of an airfoil, but at this stage, control was more important than efficiency.

He conducted his initial experiments using small-scale free-flight models, evaluating different designs. After completing these tests, he settled on a full-sized design that responded well in gusty winds, concluding that it was a practical device employing a proper method for maintaining lateral stability.[7] This glider incorporated diagonally hinged "flaps" at the outer trailing edges of the wings. The flaps were normally held in a flat position using spring tension from below and served as both load dampeners and a means to facilitate pilot-operated roll control (as is done through modern ailerons). John devised a mechanical system, consisting of a saddle and a rocker arm, to allow the pilot to alter the angle of attack of each hinged wing flap independently to help maintain "equilibrium" or lateral control.[8] This mechanism went through at least three variations, incorporating stirrups connected to a pulley through wires. With this crude system, John developed the basic control concepts that he would later refine in future aircraft designs.

Remembering the difficulties of launching the first glider by foot (or with a rope), and given the greater weight of the second glider, John completely revised his launching strategy. Through the use of a free-wheeling dolly and a trigger release, he conceived of using gravity to generate increased takeoff velocity down an inclined track. A small-scale model of the glider-launching device was tested. While crossing the Fruitland Ranch en route to the Montgomery residence during this period, neighbor Melvin Johnson unexpectedly encountered John experimenting with this device to launch small-scale glider models (wingspans of 4 and 5 feet) behind his shop, located discreetly within a grove of eucalyptus trees.[9] Upon hearing of these activities, older neighbors became even more skeptical or amused. Even Ellen felt that dabbling with flying machines was not the best use of her son's time or his potential. She chided John for these "youthful pranks," urging him to abandon his obsessive studies and rest.[10]

The Montgomerys socialized with their neighbors, notably the families of Dr. David Arnold, David Burroughs, William Blount Couts, and the family of local land-grant holder Santiago E. Argüello. Rosita Blanco Clemens had visited the Coutses as a young girl, and also received music lessons from Margaret Montgomery. Rosita later recalled John's frequent presence at the Couts home, as he and the Coutses' eldest daughter, Katie, were sweethearts.[11] Some of the teenage friends from these families would provide a source of assistance in John's experimental work with flying so long as they honored a commitment to total discretion.[12] Such assistance was limited to physical tasks such as sawing wood, sewing fabric, or hauling the glider to a suitable location and assisting with glides. One neighbor, Charles Burroughs, helped on a part-time basis with the construction of the full-sized version of the second fixed-wing glider during the winter of 1884 and into the spring of 1885. During this time, John showed Burroughs the remains

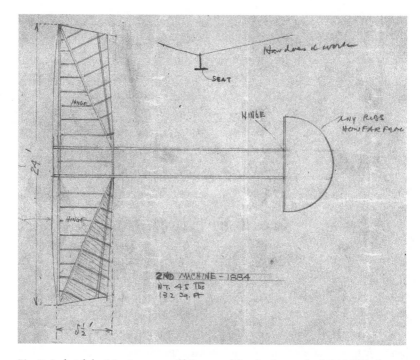

Fig. 9. A sketch by Montgomery of his second fixed-wing craft of 1885. This drawing was originally intended for inclusion in a prospective journal article (mid-
1909) but was never published. The handwritten questions (e.g., "How does this
work?") are thought to have been added by Ernest L. Jones. (Courtesy Santa Clara
University Archives and Special Collections)

of the first glider, which he had flown and crashed several months earlier.
Using an apparatus he had devised for steaming wing ribs, described as "a
box connected with his forge by pipe for the purpose of generating steam,"
John created various wing rib shapes, which were then tacked on the side
of the shop in the desired shape and left to dry.[13] He described this second
glider in a later lecture;

> This second machine had 132 sq. ft. of surface and weighed 45 lbs. . . .
> It would not carry me for anything but short jumps, even with the in-
> creased surface [area]. The machine was also provided with a mechanism
> by which I could throw the spring out. The seat, as before, was about
> 1–1/2 feet below the surface but attached thereto, was a vertical piece of
> wood which was attached by wires to the hinged portions of the wings.
> By leaning to one side or the other the level was caused to rock or move,

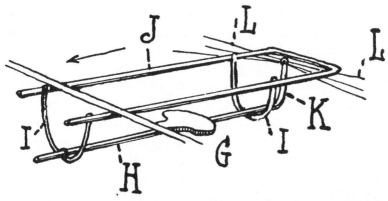

Figure IV.—Second control used on Montgomery ailerons. In this the seat G is fixed rigidly to the bar H, which rocks in the frame ribs II, which are themselves secured to the main frame J. To the upturned rear end of H, at K, are attached the groups of wires LL, one group to each wing, so that as the rider leans to one side or the other the wings are unsymmetrically manipulated.

Fig. 10. Victor Lougheed's sketch and explanation of the wing-warping control mechanism used on Montgomery's second craft of 1885. (From Victor Lougheed, *Aeroplane Designing for Amateurs* [Chicago: Reilly & Britton Co., 1912], 17–18)

pulling down one flap and allowing the other to yield. This gave me entire control of the machine in the wind, preventing it from upsetting.[14]

Burroughs later provided an affidavit for a court case (see Epilogue) confirming his assistance in late 1884 in constructing several flying machines with wingspans of 16 to 18 feet, some of which may have been mockups.[15] In addition to noting the hinged wing flaps, he recalled that the wings were braced with steel wire. John initially used waxed cloth for the surface covering, but he later obtained silk that he waxed for use in covering the wings and tail. Burroughs described a variation on John's usual control system: "He had a saddle located in the center of the machine to which was attached stirrups connected with wire leading to the tail and win[g] controller to change the angle. The tail was of a rudder combination and could be elevated or lowered by the operator."[16] Given the scarcity and expense of resources, it is likely that John reutilized the basic airframe (fuselage, operable tail, and allied hardware) from his first glider. Its general form, including an operable elevator, remained the standard design for his flying machines over the next several years.

As Montgomery and Burroughs made progress with the second glider, other events unfolded that had a major impact on the Montgomery family. In May 1885, Attorney General Augustus H. Garland submitted to President Grover Cleveland his recommendation for the position of U.S. assistant attorney general: Zachariah Montgomery of California. Garland was well acquainted with Zach's unique qualifications (particularly in terms of land patent law), having followed his career since the 1850s and having gotten to known him while they were in the graduate program in law together at St. Joseph's College in Bardstown, Kentucky. Following Garland's announcement, the Montgomerys were once again thrust into the spotlight, this time at the national level.

Just prior to the family's departure for Washington, D.C., Burroughs, Charles and Thomas Couts, and another friend named Ed Stokes assisted John in the initial experiments with a full-scale version of his second fixed-wing glider.[17] Richard Montgomery also took time away from his chores to help. The gliding location was described by Stokes as "a small ravine just east of Wheeler Hill."[18] Here they conducted initial test flights. Mary F. McCarthy, who served as a live-in housekeeper at the Montgomery home from 1882 to July 1885, recalled that John discussed his progress with his siblings: "I remember him coming in, you know, on several occasions, where he had made trips, and heard him talking about how many feet he went up, you know, and how he came down, and all of that, but of course I was young and did not take much interest in it."[19]

On July 4, 1885, Zach, Ellen, Mary, and Jane left for Washington, D.C. Only John and Richard remained at Fruitland, and they found themselves having to quickly adjust to their newfound bachelorhood. Some months later, John sent a playful description to his sister Margaret:

> But cast your eyes westward and behold your two bachelor brothers living on this large and lonely ranch. You should have a peep at us and our surroundings; you know Dick and I are very fond of keeping a house in order and [you] may judge for yourself how things are in this large place, Dick being engaged in farming and I at experimenting. I do most of the cooking; but Dick sometimes takes compassion on himself and wants a change; I very politely step aside, while Dick gets the cook book; then I come in for my share of good living.[20]

The family's departure was reported by the local press in San Diego, which also mentioned John's inventiveness: "The eldest son, John J. has won some distinction as the discoverer of a process for de-vulcanizing rubber."[21]

With the family now gone, John was free to conduct gliding experiments without raising his parents' concern. He decided to find another location for gliding that would be even more private. The eastern part of Otay Valley lay far from any settlements and was essentially undeveloped. He had already become familiar with this particular area near the San Ysidro Mountains (specifically Otay Mountain) while collecting his eaglet and observing birds soaring adjacent to a steep bluff.[22] Richard later recalled that in 1885 he and John identified a canyon located on the north side of Otay Valley, about eight to ten miles east of Fruitland Ranch.[23] After clearing the brush from the hillside, they constructed a 50-foot-long elevated, wood-framed glider-launching track that followed the slope of the hillside facing the direction of the prevailing wind (facing west). Richard recalled that the ramp "was quite high and very abrupt . . . it was built out of 2 by 3's and covered with steel strips on top . . . two by three scantling; and then bolted crosswise to hold it a uniform distance below the two parallel tracks, the same as a railroad track. It was anchored very thoroughly to the top of the hill." The glider rested on two supports on the cart, and it was trigger-released through a mechanism located near the end of the track. The cart was stopped at the end of its run by an anchor chain, and could be drawn back to the top of the track by a rope.[24] A review of primary accounts (as well as comments by Ed Stokes and others made at a later time), coupled with historic maps, wagon routes, aerial photos from 1928, and prevailing wind direction, strongly suggests that this gliding location was on a ridge that tended from north to south, located on the north side of Otay Valley, just southwest of the present Lower Otay Reservoir. Richard participated in the initial trip to this new location and assisted with an unmanned launch of the second glider. It was released and glided a distance that he later estimated at between 100 and 200 feet.[25] Charles and Thomas Couts, Ed Stokes, and Charles Burroughs assisted with other launches at this location. Stokes described short hops of 50 feet in length, and Burroughs witnessed a number of glides of between 200 and 300 feet. On the basis of these initial trials, John was pleased with the controls:

> I found that when I took that machine and faced the wind that its equilibrium was perfect, that is, I found no conditions under which I could

not control it so it would not upset me, a thing I could not do with the first machine. When I attempted to glide I found its power of gliding was far inferior to that of the first. Immediately I found that I had not found the right surface.[26]

After the family had moved to Washington, D.C., John and Richard occasionally wrote to keep them apprised of life at the ranch.[27] Near the end of July 1885, Richard wrote to Zach specifically, noting John's initial success with his second flying machine at the east end of Otay Valley. Zach's optimism about his son's eccentric flying activities was evident in his playful response: "We are now waiting anxiously to hear the result of his next experiment with his flying machine. The girls say they will watch for him to land on top of the Washington Monument."[28]

Yet John's goal of emulating the birds remained elusive: "In my crude attempts at imitation," he said, "under identical conditions, my gliding machine would make only short flights and gradually descend. And it not unfrequently [*sic*] happened, while I was floundering along in my machine, these birds would glide by time and again, or circle around, and ascending in bounds, reach elevations of hundreds of feet in a few moments."[29]

John remained convinced that the laws of fluid mechanics applied generally to both liquids and air. He commenced a series of experiments in controlled conditions to better understand the movements of particles around different surfaces.[30] This methodical, scientific approach set him apart from many others who were working on the problem of flight at this early time. With the benefit of hindsight, and in reference to men such as Montgomery and Samuel P. Langley, years later Octave Chanute noted, "about the year 1885, the problem of aerial navigation began to be embraced in the United States, by a number of men who were not quacks."[31]

To study his theories in a controlled setting, John designed a whirling-arm table to measure the lift generated by the surfaces of different curved shapes.[32] Similar devices were developed independently by other aviation pioneers, including Horatio Phillips (England), Otto Lilienthal (Germany), and later Samuel Langley (United States). Montgomery's whirling arm consisted of two parallel rails fastened together and mounted on a pivot. On the end of this he fastened surfaces of different forms to study their movements against wind. He quickly observed that there was something happening to the flow of air taking place in front of each airfoil shape.[33] To further explore this phenomenon, he took a large barn door into a field near Fruitland, where, with the assistance of his brother Richard, he conducted some experiments, noting the manner in which large particles such as light down moved around the door in a prevailing breeze with the door held at various angles.[34] He threw thistledown into the wind and observed that upon nearing the barn door, which was being held at an angle by Richard some distance away, the down turned, approached the surface, and then glided around it. It then continued in a long, wavelike curve resembling what John described as "the floating of an immense flag." This phenomenon evoked several fundamental questions for John: "What was the cause of these movements? What did the surface shape have to do with them? What was the nature of the collision between the moving air and the surface? Was it elastic impact or simple static pressure?"[35]

John constructed a crude wind tunnel to investigate these same phenomena under controlled conditions.[36] He introduced cigar smoke into an inlet and used a heliostat to illuminate the currents as the air interacted with small airfoils.[37] Although the first documented use of a wind tunnel was by Francis Wenham in England in 1871, crude wind tunnel devices were also used by some of Montgomery's contemporaries, including Lawrence Hargrave (Australia), Horatio Phillips (England), and Otto and Gustav Lilienthal (Germany). John converted the wind tunnel into a water tank in which currents could be generated and regulated through hydraulic systems. He placed metal surfaces in the tank at various angles to observe their interaction and the distribution and relative movements of pressures. The movements were observed by means of dust scattered on the surface, and with a silk fiber attached to a slender wire and used as a flag. The dust revealed the general motions, while the flag indicated the direction of motion in any particular portion of the fluid. John noted that the great difference in elasticity of the two media (air and water) had no appreciable effect on the general results.[38]

Through these experiments, John identified the location of pressures that developed on the surfaces when they interacted with a current. He noted that the center of pressure changed as the angle of attack of the surface varied, and he confirmed that camber-shaped surfaces generated the most lift. With these observations, as well as his work with practical models and manned gliders in 1883–85, he concluded that a curved airfoil was necessary for the production of lift. The superior lifting capability of what some experimenters referred to as "concavo-convex" surfaces relative to planar airfoils was confirmed independently in the 1880s by a handful of Europeans, including Horatio Phillips and Alexandre Goupil.

John occasionally sought to distract himself from his obsessive studies.

Fig. 11. Montgomery's water-current tank, for testing the interaction of small air-foils with particulate matter in water currents. It was modified (ca. 1885) from an earlier wind tunnel for testing the interaction of small airfoils with smoke in air currents. (Courtesy Santa Clara University Archives and Special Collections)

Some youngsters on neighboring ranches had formed the Otay Literary Society and met regularly to review and discuss the latest literature of the day. John served as society president and enjoyed the social camaraderie. A family friend, Rosita Clemens, recalled an incident in 1885 that illustrated John's penchant for occasional pranks. At the end of each society meeting at Fruitland, the women members, feeling sympathetic toward their bachelor hosts, would insist on cleaning up after the gathering. But John, not want-ing to be pitied, would protest to no avail. For the next gathering, he devised a little surprise: he rigged up a small battery in his room and ran wires to the wash basin in the kitchen:

> Next time they came and I reminded them about the dish washing. . . .
> I fixed matters so that whoever attempted to wash the dishes would
> get all the electricity she wanted. She no sooner put her hand in the water
> than she jumped back, and with a most terrified look, pointed in a tragic
> manner at the pan, and said "something is the matter with that water, I
> declare there is—oh my arms!" You never saw a more astonished set of

persons: but they merely all accused her of "putting on" but undeceived themselves when they tried to take the cloth out of the water.[39]

In December 1885, John wrote to his sister Margaret, who was serving as a nun at the convent of St. Joseph's of Carondelet at St. Louis, to update her on his progress:

> My attention is still fixed on my flying machine and I am getting along fast enough to suit myself. Since I last wrote you, I have performed hundreds of experiments and discovered some important facts and laws. I have had many failures and discouragements but I have become more convinced than ever of the correctness of my ideas and plans. My theory from the beginning has not changed, though at first it was not by any means as clear as now, but the difficulty I have encountered is to combine the proper mechanical principles to work out this theory. . . . I consider that providence has been very gracious with me; for I have acquired invaluable information on the subject, and indeed so strange is it in its nature that I will have to *prove* the truth of my information by the success of my machine before anybody having a knowledge of Philosophy will believe me.[40]

Fully aware of the risk he was taking in experimenting with flying machines, he closed with a comment that would prove to be telling: "I wish you to *always* pray very earnestly for my *safety* and success."[41]

Even at this early date, John understood the importance of demonstrating his ideas in front of qualified scientists in order to confirm their validity. Such a demonstration would eventually be required for the protection of intellectual property as well. Inventors were unable to secure a patent until they could demonstrate the practical performance of their invention. In a letter to Chanute, John's contemporary Louis Mouillard identified secrecy itself as a major problem: "The greatest danger to aviation is the spirit of mystery in which those who take up the study of aeronautics generally cloak themselves."[42] A few years later, Chanute argued for the need for others to protect inventors' interests by maintaining silence until concepts were patented.[43]

Having eliminated the flat airfoil as a proper model, John reviewed the variety of curved airfoils represented by birds' wings in the surrounding area. In particular, he focused on the turkey vulture (*Cathartes aura*), which he referred to as a "buzzard." In a letter to Margaret, he indicated that he was building a large flying machine with which to test certain facts developed while he was carrying out a series of experiments during the

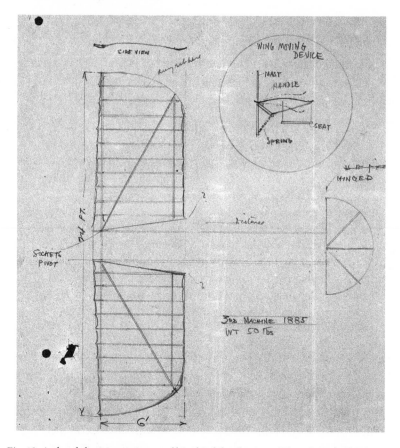

Fig. 12. A sketch by Montgomery of his third fixed-wing glider of 1886. This drawing was originally intended to be included in a prospective journal article (1908–1909) but was never published. The handwritten questions are thought to have been added by Ernest L. Jones. (Courtesy Santa Clara University Archives and Special Collections)

previous few months.[44] This new airfoil became the main design feature of his third glider, and construction began in December 1885. Like each of his previous designs, this third design was preceded by experimentation with small-scale free-flight models. John recalled in 1910:

> In this machine, in a way, I copied nature in regard to surface and in a way I departed from it. The wings were formed more or less like those of a soaring vulture, with this exception: I could not bring myself to the belief that the surface curved down in the front was the proper surface. Therefore, I compromised by turning the front edge up a little and the

rest of the wing was more or less like that of the vulture. The two wings were placed at a dihedral angle. Now in this machine I carried out the warping principle in a different way. There was a lateral beam along the front of the wings. These two beams were capable of being rotated in a socket in the frame extending backward to the tail. Wires from the rear of each wing ran to levers, one for each wing, placed at the right and left hands of the operator, who sat on a seat as in the other machines. By these levers I could bring both wings down together, or independently. That machine was perfect in control; whether the wind was regular or gusty I had the machine under control by changing the angles of the wings. This had larger surface even than the second but was inferior in lifting power. Immediately I found I did not have the proper form of surface as it did not have the same lifting power under the same conditions as the first machine.[45]

By setting the wings at a dihedral angle, thereby mimicking the turkey vulture in soaring flight, he increased the lateral stability. He continued to address roll control by varying the angle of attack of the wings (either in unison or independently) while in flight.[46] However, the wings (when operated in unison) also addressed pitch, as the elevator in this craft was fixed to control pitch. Only very short glides with the third glider were accomplished at the rail-launch facility; John considered its lifting capacity to be poor.

Through his understanding of ornithology, John was the first person in the western United States to study the mechanics of bird flight. He was a pioneer in America in using small-scale free-flight models and other devices to test his aeronautical hypotheses and evaluate successful designs through man-carrying flying machines, all toward a better understanding of lift and flight control. His isolation at Otay was compounded by a lack of resources, societal pressure, family obligations, his father's political career, and a desire to preserve his intellectual property in the discipline. As James once commented, "John believed that, through adversity grew fortuitous circumstances."[47] By 1886 at Otay, John recognized not only that controlled human flight was possible, but that a parabolic airfoil was a requirement for success.

CHAPTER 6

The Path to Recognition

I suppose the folks all believe a little in my ideas now.

John J. Montgomery to James Montgomery, 1886

As the process for confirming Zach as the nation's assistant attorney general began in the late spring of 1885, a new controversy emerged. Zach's former political opponent Stephen J. Field was a member of the Senate Judiciary Committee and thus represented a key vote in the confirmation.[1] Justice Field and Republican senators John J. Ingalls of Kansas and George F. Edmunds of Vermont were staunch proponents of the "Blair Bill" (Senate Bill 194), an education appropriations measure authored by Senator Henry W. Blair of New Hampshire. Field, Ingalls, and Edmunds theorized that because the appropriations for the bill would be administered through the Interior Department, Zach would surely mount a challenge through the Office of the Attorney General. In an effort to undermine his confirmation, therefore, the Judiciary Committee brought his stance on "the school question" to the fore during the review process. Advocates of the bill circulated an abstract of one of Zach's more famous San Francisco lectures on public education in an attempt to misrepresent his views to the committee.[2] Augustus Garland announced that he would resign his post as attorney general if Zach Montgomery was not confirmed.[3] A yearlong battle ensued.

While the fight was being waged in the Senate, the nation's leading newspapers followed the story closely.[4] Zach was often represented as an enemy of education, especially by the press on the East Coast. His appointment was rejected twice by the Senate Judiciary Committee.[5] However, President Cleveland interviewed both Zach and Garland, read Zach's speeches that had fueled the debate, and chose to defend him. In the spring of 1886, he held private meetings with the committee members and resolved the situation personally.[6] As Zach later wrote, "In this, Mr. Cleveland proved himself a man of remarkable independence and courage, and placed the writer under a debt of gratitude that he can never forget."[7] On April 22, 1886, nearly a year after assuming office, Zach was finally confirmed as assistant attorney general.

James was living with the family in Washington while he attended Georgetown University.[8] As he browsed the newspapers one morning, he noticed an article written by Israel Lancaster titled "The Soaring Birds." It highlighted a speech that Lancaster had delivered on August 18, 1886, to the American Association for the Advancement of Science (AAAS) on the flight of buzzards along the coast of southern Florida.[9] James forwarded the article to John, pointing out in a cover letter that Lancaster might be considering patenting some of the ideas presented in the speech. At the time, John was working as a deputy collector and customs agent at the customs house at Tia Juana, California, near the border with Mexico (at present-day San Ysidro). He had been recommended for the post by his brother Richard, who had been forced to take a leave of absence from the position because of a back injury.

Given the similarity of Lancaster's work to his own pursuits, John found Lancaster's commentary on soaring flight unnerving. The lecture had a similar impact on other aviation pioneers in that period, most notably Samuel Langley.[10] Concerned that he would be surpassed in the pursuit of invention, John ended his short stint at the customs house and moved quickly to determine just how far Lancaster had progressed with patent protection. In early September, he wrote to ask James to contact the AAAS and find out how John might present a paper on his own work, thereby establishing prior art. Additionally, he directed James to ascertain whether Lancaster had filed a patent application. John felt compelled to discuss his work as a scientist, but he remained cautious (even anxious) when it came to revealing that work to others. In closing the letter, he expressed his concern about being usurped: "If I should be beaten in my problem—it would about kill me!"[11]

Lancaster's speech was greeted with a decidedly negative reaction in the press. Said one newspaper article, "Some of the members of the Association seem to be in a quandary as to whether Mr. Lancaster is a crank, or a sharp practical joker who has been giving the great association of America's savants guff."[12] Octave Chanute later contended that Lancaster had failed to adequately explain his theories at the Buffalo conference: "Unfortunately for Mr. Lancaster, upon the latter occasion he attempted to give a mechanical and mathematical explanation of the performances which he had previously so well described, and his theory was so plainly erroneous that he was subjected to hard ridicule and criticism."[13] The intense scorn directed toward Lancaster was a clear indication that the public was nowhere near ready to accept the notion of manned heavier-than-air flight. Despite his significant interest in the topic, even Chanute remained reluctant to publicly associate himself with the early aviators, afraid of losing his excellent reputation as a railroad engineer.[14]

Shortly after receiving John's letter, James visited the U.S. Patent Office to research prior art in heavier-than-air flying machines. Admittedly out of his depth in such an investigation, he discovered various patents and effigies of flying machine designs, but nothing similar to John's invention, nor any evidence suggesting that Lancaster had applied for a patent. After learning of James's reconnaissance, John concluded that he still had sufficient time to refine his airfoil and aircraft design through experimentation. He held off on filing a patent application or revealing his work through a lecture.

It was not easy at the time to convince the Patent Office that a successful man-carrying, heavier-than-air flying machine was even possible. As late as 1893 and 1894, aviation pioneers Mouillard, Chanute, and Hiram Maxim were still having difficulty making a case for their patents. One patent examiner reacted to Mouillard's patent application of 1892 by suggesting that "As a whole it is not practical since the machine cannot rise without a balloon." He also refused to rescind his objection "unless it is a model in full operation which would be capable to make an ascent and to be controlled."[15] Upon hearing this, Mouillard noted to Chanute: "I think that without being wicked, one may say that they have not studied the aviation problem sufficiently. On the one hand the Institute, Mr. Langley, and you yourself and even nature, on the other hand 'without a balloon the machine cannot rise,' it does not make any sense. . . . If the facts were practically demonstrated that would be all that is necessary, the demonstration would be rather important and rather obvious to give the entire matter an absolute certainty of propriety."[16] In fact, in 1893, Mouillard considered submitting an affidavit describing his earlier glider flight with his *Model No. 3* at Mitidja, Algeria, to establish to the Patent Office that flight with a heavier-than-air flying machine had already been accomplished without the assistance of balloons. As Mouillard specifically conducted his experiments in private, without witnesses or photographs, and offered no statements to the press at the time of his gliding flight in 1865, his only means of authentication was through an affidavit. Upon further reflection, he decided that submitting an affidavit was futile. In arguing for the value of Mouillard's affidavit in 1893, Chanute remarked, "Mr. Montgomery of California has made almost an identical experiment."[17] Given this state of affairs, one may wonder which was more foolhardy: the attempts of early inventors to develop flying machines despite ridicule and isolation, or the closed-mindedness of the U.S. Patent Office when considering patents on heavier-than-air flying machines.

In late 1886 or early 1887, John traveled to Oakland to visit his aunt Margaret McCourtney and some old friends. During this trip he met Lander Redman.[18] Formally educated in mathematics and law, Redman recalled a long conversation they shared one evening at the McCourtney home in 1887. Among other topics, John discussed his gliding experiments at Otay, aeronautical theories, and research in astronomy: He

> had flown himself, down I think, in San Diego County or possibly in Lower California, where he had been living, and I remember his telling me about having been injured in one of his flights. I don't recall now the distance that he said he had flown in his aeroplane, but some considerable distance, and he went quite elaborately into an explanation of the theory of flying. He spent a long time explaining to me the reason that the soaring bird did not fall.[19]

Redman was intensely interested in astronomy, especially planetary motion. He and Montgomery corresponded regularly over the following months, and John offered elaborate descriptions of his theories on the subject.[20] When asked later, "What was his standing in the scientific world, as far as you know?" Redman responded:

> I can only answer to my own estimate of the man. . . . I was just like a child in the hands of a person of extraordinary capacity—a genius shone from the man's brain and his intellect was as sharp as a cambric needle. There was not a phase of the matters that we talked about that night that he did not have at his fingertips and was not able, instantly, clearly and forcibly to explain.[21]

In articles published years later, Redman offered insight into John's theories of planetary inversion and how they accounted for the precession of equinoxes, the tides, the transfixed position of the moons of Jupiter, Earth, and Venus, the tilted axis of Uranus, the retrograde rotation of Neptune, and the motion of Saturn's ninth moon.[22] Frustrated when Professor William H. Pickering received credit for John's theory of planetary inversion in 1905, Redman wrote that Montgomery's "self-abnegation is so extreme that unless someone else drew attention to his claims the merit of the

Fig. 13. The Montgomery family at Fruitland in 1889 as John prepared to take Mary to the railroad depot for the trip to Washington, D.C. Left to right: James, John, Jane, Mary, Ellen, Zach, and an unidentified household helper (seated). (Courtesy Santa Clara University Archives and Special Collections)

discovery referred to would naturally be bestowed wholly upon Professor Pickering."[23]

In the fall of 1888, President Grover Cleveland was defeated by Benjamin Harrison in his bid for re-election. His loss was a bitter pill for Zach to swallow. Zach was disenchanted with Washington, D.C., and by those within his own Democratic Party who he felt had worked against Cleveland during the campaign. In March 1889, only three months into Harrison's term, Zach resigned his post. Three months later, following James's graduation from Georgetown, the family returned to Fruitland. Margaret also came back after time spent at the convent of St. Joseph's of Carondelet at St. Louis. Zach reopened a law office in San Diego, and in 1890 he entered into a partnership with James, doing business as Z. Montgomery & Son. The reconsolidation of the large Montgomery family at Fruitland brought visits from old acquaintances and relatives.

John's obsessive nature led him to spend long hours in the lab studying and experimenting. He occasionally set aside his work in aeronautics to continue studies in electricity and astronomy. Irritated with the interruptions from the occasional visitors to the ranch wishing to buy fruit from the extensive orchards, he hung a sign on the entrance to the barn: "Free Fruit—Help Yourself."[24] Science was his passion, and it was hard for him to tear himself away. But he was soon to have a respite from his intensive labors—a break that would take him far away from Otay and place him on the path to recognition.

CHAPTER 7

Chicago, a Forum
for the Outsider

oncerned about the effect that John's obsessive studies were having
on his health, his siblings urged him to take a vacation. He decided
to take a train trip to attend the 1893 World's Columbian Exposition
in Chicago. Although the Montgomery family finances had been hit hard
by the failure of the California Savings Bank of San Diego during the reces-
sion of 1893, Zach pulled together enough money to fund his son's travels.
John's visit to Chicago was timed perfectly with two significant events on
the agenda: an International Conference on Aerial Navigation, scheduled
for August 1–4, and a lecture on electricity to be given on August 25 by
Nikola Tesla, a Croatian-born genius who conducted visionary work in
electricity, including dynamos, alternating current, and other inventions of
direct interest to John.

As the first of its kind in the United States and only the third in history,
the aerial navigation conference brought together leading authorities on the
nature of flight and its potential use by man. Albert Zahm, a young aeronautical engineer and professor of mathematics at the University of Notre
Dame, served as the general secretary of the conference. Octave Chanute
was initially concerned that the meeting would attract "cranks" rather than
scientists and engineers, and he was skeptical about associating his name
with the venture. It was only through a great deal of persuasion and planning by Zahm that Chanute eventually accepted, with "unfeigned trepida-
tion," the role of conference chairman. Zahm recalled: "Though he suffered
not a little raillery from his brother engineers for his interest in an 'impossible' art, he pointed with some elation to the forty-five papers presented by
the industrious members of this congress."[1]

To attract high-quality researchers, the aerial navigation conference
was purposely scheduled to coincide with an engineering conference, also
to be held August 1–4.[2] In preparation for the event, Zahm and Chanute
assembled papers on every topic of aeronautics, to be given by writers and

experimenters whom they regarded as foremost in that science throughout the world.[3] After arriving at the exposition, John made his way to the engineering headquarters and introduced himself to Zahm. As Zahm recalled, "After a short conversation, I introduced him to Chanute who had just finished writing his opening address as chairman. We all became friends, and Montgomery agreed to tell the conference about his experiments."[4] Given Chanute's initial skepticism about the conference, it is all the more remarkable that John so quickly earned their confidence.

As it turns out, John and Chanute had a point of common interest even before the conference. In March 1892, Chanute had visited San Diego, where he had studied the flight of birds from the bridges of steamers docked at the wharf.[5] Upon returning to Chicago, he wrote to Mouillard on June 17, reporting his thoughts and conclusions in great detail.[6] He would return sporadically to San Diego over the next decade or so to make further observations of bird flight. Chanute not only made some very careful and detailed observations about the flight characteristics of birds, chiefly their ability to soar, but he also made a logical connection between the different airfoils used by flightless and flying birds as a direct result of his observations at San Diego. In company with assistants and friends, he produced many instantaneous photographs of pelicans; he was so impressed with these interesting and picturesque birds that he commented to an assistant that the secret of air navigation lay in imitating their flight. Hundreds of photographs were taken, to be studied later in his Chicago laboratory.[7] He also visited the U.S. Weather Service while in San Diego, where he and meteorologist Ford Ashman Carpenter discussed their interest in both weather and photography.[8]

Illinois aviation historian Howard Scamehorn notes that in his opening address to the aerial navigation conference, Chanute "notified the assembled delegates and the world that the members would consider only the work of practical, serious men seeking the advancement of flight through scientific experiment."[9] He strongly believed that the results of aeronautical research should be shared freely among the scientists engaged in solving the problem. While patenting was possible, and indeed encouraged, the patent process was intended only to preserve inventorship. The results themselves, he felt, should be made available to the community without compensation.

For the first time in his life, John found himself surrounded by other engineers and scientists involved in the full breadth of aeronautical research. He arrived at the lecture hall on the first day just in time to hear the reading of Samuel Langley's paper on "The Internal Work of the Wind." John felt that Langley's observations generally agreed with his own, a fact that gave him "a feeling of greater assurance, a sort of moral support, for there is a certain degree of comfort to be got out of having someone else see things as you do." He added: "He had observed certain facts in relation to the wind, and I found out the reason for them, so the paper was brought up again for discussion. I felt that my theory was the key to the problem of aerial navigation."[10] John's first lecture was given August 1 or 2 and focused on his laboratory experimentation with airfoil shapes in air and water currents, as well as his anatomical investigations of birds, specifically bird wings, made during 1885–86. Shortly thereafter, he wrote about his experience to his sister Mary: "It's a grand thing to be among the leading scientists of the world, being introduced to them as 'a young man who, cut off from all scientific association, has for years through pure love of science devoted himself to scientific investigation in its most advanced stages,' and then to be met with the remark 'If I can do anything to assist you, let me know.'"[11]

Chanute and Montgomery spent long hours conversing at Chanute's home in Chicago. As each revealed more to the other about his own work in controlled flight, John gradually shared details about the methods he had used in his craft of 1885 and 1886 and the fact that he wanted to patent this technology. Through these conversations, Chanute inspired a change in John's thinking toward patent protection. John later wrote: "we agreed that no man making such a discovery had a right to monopolize it, because it was not like the discovery of some new commercial utility of limited value, but was a step in the progress of humanity and therefore belonged to humanity."[12]

With Chanute's encouragement, John decided to disclose the nature of his inventions as part of a second lecture, delivered sometime during the last two days of the conference.[13] This was precisely what he had briefly considered doing for the American Association for the Advancement of Science in 1886. This second lecture was preceded by a very brief preamble in which he described his early gliders and gliding experiments of the early 1880s at Otay. The main body of the talk, however, focused on his methods and theories for the maintenance of equilibrium in flying machines. His peers learned that he had completed multiple gliding flights testing a series of designs, as a result of which he had developed automatic and pilot-operated control mechanisms for lateral and longitudinal control. Zahm later confirmed that during this second lecture, Montgomery "told us he used oblique flaps for lateral control."[14] Years later, John commented in hindsight: "I was working purely as a scientist, with no intention of making money and I proposed publishing my discoveries for all investigators, and giving it to the world as I did, I did not think it necessary to take out patents in those circumstances."[15]

After the conference on aerial navigation, John stayed in Chicago through the end of August in order to participate in the World's International Electrical Congress, where he became a member and attended Tesla's much-anticipated lecture, which was titled "Mechanical and Electrical Oscillators." Afterward he traveled to Washington, D.C., to visit Mary, then returned to California by way of the San Francisco Bay area. He stopped in San Francisco to visit an old friend and former St. Ignatius College classmate, Francis J. Cleary, at the Cleary family's residence on Howard Street. Frank's younger sister Regina already knew a little bit about John: "I [had] heard Frank speak of him. I [initially] confused John with Zach, his father."[16] During the visit, nineteen-year-old Regina listened in delight as John recalled his time at the World's Fair and the way he had been received by the leaders of the conference. He also spoke of his life in San Diego. In an interview in 1946, Regina revealed that it was during this visit that she first developed a love for John.[17] Many years passed before she discovered that during the same visit, John had also fallen in love with her. At the time, however, John's circumstances were such that he did not feel he could provide for a wife, much less start a family. Their love would remain unrequited for a very long time. Regina later commented on John's decision to forgo a marriage proposal in 1893: "he was in love with me for 15 years—rather than put me through the hardships he had seen."[18]

In the mid- to late fall of 1893, John responded to a request from Chanute by sending him a concise description of his earliest gliders and experiments in the 1880s, which Chanute forwarded to Zahm for potential inclusion in the conference proceedings.[19] Because he was naturally cautious, and perhaps because he knew of Chanute's tendency to assimilate and redistribute aeronautical knowledge, John's brief communication lacked sufficient detail to enable anyone to faithfully re-create his intellectual property. When writing an article for the series "Progress in Flying Machines" in the *American Engineer and Railroad Journal,* Chanute drew his discussion of Montgomery from the unedited stenographer's notes covering Montgomery's second speech as Montgomery's original speech summary was not available (it was with Zahm). Although some historians have assumed that Chanute's resulting published article was a primary account, it was in fact his interpretations of the unedited stenographer's notes from Montgomery's speech.

Meanwhile, John's first lecture on his wind tunnel and water-current tank experiments and his ornithological observations in San Diego was edited by Chanute and published in the July 1894 issue of *Aeronautics* with the title "Discussion on the Various Papers on Soaring Flight."

John had also expected to see his second speech at the conference in print. Chanute had retained the stenographer's record of that presentation, and John assumed that it would be included in the conference proceedings being compiled by Chanute and Zahm for the *American Engineer and Railroad Journal.* Perhaps because of John's shift concerning disclosure of detail about his technology, some confusion occurred between the two men. Correspondence between John and Chanute some months later reveals that Chanute omitted the main body of the second lecture from the published proceedings under the mistaken belief that it would reveal detail and technology that John was intending to patent. When John asked in the spring of 1895 why it had not been included, Chanute responded in part: "I am puzzled by your reference to a paper written by you in Chicago which was not published. You did give me a record of your experiments but I thought distinctly that you did not want them published."[20] Further correspondence indicates that Chanute asked Zahm to return the stenographer's record of the speech to John in the summer of 1895. In 1910, John reprised the speech in an appearance before the Aeronautical Society of New York, and at that time it was excerpted in various articles and aeronautical serials.[21] Even though this second speech was not published at the time, John's public disclosure of his inventions at the conference was an important factor in the later establishment of due diligence.

In 1893, the Jesuit fathers at Mount St. Joseph's College at Alton, California, began a search for a new science instructor. Father Richard H. Bell, a Jesuit physicist and a former classmate of John's at St. Ignatius, recommended him for the position. John accepted the offer in March 1894 and relocated from Otay to Alton.[22] Alton was a small hamlet in Humboldt County, on the sparsely populated coast of northern California, 250 miles northwest of Sacramento. In many ways it was even more isolated than Otay. In accordance with Jesuit philosophy, Mount St. Joseph's College was a closed campus, a policy that was designed to create a nurturing, intellectual environment for the students while minimizing outside societal influences. These constraints further hampered John's opportunities to experiment with manned gliders. After the Chicago conference, Chanute and Zahm showed a keen interest in his experimental work, and it was with their encouragement that he undertook a new round of experiments in 1894, using a wind tunnel and water-current tank at Alton to investigate the currents moving around surfaces so as to determine a theoretical and experimental basis for the proper airfoil.[23] In the winter of 1894–95, he summarized his airfoil theory in an unpublished manuscript titled "Soaring Flight."

His prior investigations with a wind tunnel and water-current tanks in the 1880s had confirmed the existence of relative distributions and movements of pressures on an airfoil placed at various angles to a current, in addition to the existence of an area of low pressure (or what he termed "rarefaction") near the top leading edge of an airfoil. However, his attempts to more thoroughly investigate these phenomena had been hampered by the primitive state of his scientific equipment.[24] This new phase of experimentation was much more fruitful and culminated in a model of the flow field around an airfoil section and the forces produced by the airflow on the surfaces. With this he was able to solve the dilemma that his earlier airfoils had presented. As he noted a decade later, "It required the nicest precision in mathematical calculation, of course—in conical sections chiefly. . . . I worked it all out 10 years ago and set it aside."[25]

Recent independent evaluations of Montgomery's "Soaring Flight" by three aerodynamicists (Dr. Mark Ardema, Dr. John H. McMasters, and Dr. Gary Flandro) suggests that Montgomery devised a correct theoretical model for the generation of lift (what Flandro has termed "Montgomery's Lift Mechanism"), in addition to having attained an understanding of airfoil efficiency and stability.[26] In the late 1800s, prevailing theories on the nature of flow over an airfoil were in conflict regarding models resulting in the generation of lift and drag. This result puzzled researchers for a very long time.[27] An important insight from Montgomery's model is the presence of a circulating flow superimposed onto the streamlines on the airfoil. This demonstrated his understanding of the significance of vorticity (or "re-circulating, rotating flow component") in the production of lift.[28] The model allowed the flow to better represent the observed experimental behavior of an airfoil. However, Montgomery's circulation theory was contrary to the popular opinion of his peers at the time.

In reference to Montgomery's flow model at the conclusion of "Soaring Flight," McMasters noted, "He's so close here to what actually happens on a curved, flat-plate airfoil, it makes me want to grab him by the throat, and say: 'Yes, and then what?' The problem here is that he's gotten a lot of the flow physics right (but not quite all because of serious problems with his experimental apparatus), but he then stops short of synthesizing his results into some sort of quantitative formula or rule that would let one figure out how much lift one could generate with a wing of a given size—thus provided the necessary guidance to a designer."[29] Although Montgomery devised a model of the flow around an airfoil and understood aerodynamics, there is no direct evidence or surviving documentation that he derived a quantitative formula for actually calculating the lift generated by a wing surface.

One of John's contemporaries working in England, Frederick W. Lanchester, independently developed a preliminary circulation theory of lift in 1894–96, but his resulting paper of 1897 was rejected for publication by the Physical Society of London.[30] Soon after the turn of the century, in 1902 and 1906, respectively, Wilhelm Kutta and Nikolai Joukowski (Zhukovsky), also working independently, used a circulatory airflow model to which they applied mathematics to calculate the lift of a wing. Their approach subsequently became known as the Kutta-Joukowski theorem. Aviation historians generally credit Lanchester with being the first to grasp the role of the trailing vortices behind lifting wings and as the initiator of the circulation theory of lift through his published books *Aeronautics* (1907) and *Aerodonetics* (1908). However, working from Lanchester's published concept, Ludwig Prandtl published (in 1918–19) what many consider to be the first full physical understanding of the production of lift by a wing, which would come to be known as the lifting-line theory, or Lanchester-Prandtl wing theory.[31] Despite these independent successes, it should be noted that Montgomery's circulation theory predated the studies of all of these other researchers and was published in various forms between 1897 and 1905.[32]

Chanute was particularly inspired by the practical flying experiments of John Montgomery and Otto Lilienthal. He realized that these scientific men also had the rare engineering aptitude and skill to translate aeronautical theory into practice. In January 1894, when Zahm substituted for Chanute at a lecture at the Franklin Institute in Philadelphia, Chanute suggested that Zahm be sure to include Montgomery as one of a handful of men who had actually glided in the air, adding, "they have achieved nearly equal success with the passive balloonists, for they have sustained themselves in the air."[33] Soon after the end of the 1894 congress, Chanute decided to ask John to assist him with some gliding experiments. In March, when Chanute's wife and daughter traveled to San Diego by train for a vacation, he asked them to call upon John at Fruitland to extend a formal offer. By this time, however, John was already at Mount St. Joseph's College at Alton. Arriving unannounced at Fruitland, the Chanutes met Ellen and Margaret. Margaret later recalled, "my mother said she refused to tell them [John's whereabouts], because she was always so afraid about his flying; she was nervous about John; and she did not want him to begin manufacturing again."[34] Thwarted by Ellen's uncooperativeness, Chanute wrote directly to John on March 30, with a formal invitation to collaborate with him in gliding experiments. He offered to come to San Diego, or as an alternative, suggested that John was also welcome to join him in Illinois.[35]

It was also in 1894 that Chanute published a landmark compilation in the history of aviation. *Progress in Flying Machines* highlighted what he considered to be the most relevant work in aeronautics.[36] The primary purpose of the book was to rectify the general state of ignorance among enthusiasts on the "state of the art" in aeronautics by revealing the breadth of what had been accomplished in the preceding decades. The experimenter could thus ascertain which technology and approaches had already led to dead ends, and which proposals offered promise. In the book's conclusion, Chanute singled out four experimenters whom he considered to have provided the best data for use in applying motive power to heavier-than-air flying machines: Mouillard, Lilienthal, Le Bris, and Montgomery. His recognition of the value of Montgomery's work was a point that has not been lost on modern aviation historians. It was clear that he considered the manned glider to offer the surest route to a successful powered airplane. As modern aviation historian Tom Crouch noted, "Significantly, the men in whose work he took the greatest interest were gliding pioneers like Mouillard and Montgomery."[37] With the passage of time, it would come to be viewed as a classic of aviation literature and the earliest published compilation that enabled other experimenters to ascertain the state of the art in aeronautical research.

In August 1894, Chanute once again repeated his offer to collaborate with John, and this time he noted his willingness to provide financial support for the experimentation as well as exposure to the scientific community. John eventually received Chanute's proposal, but he chose to turn it down. In a letter to his sister Mary at the time of the Chicago conference, he had revealed that he was wary of collaboration with others. He felt that he was working ahead of his contemporaries: "I am convinced more than ever that it is better for me, while being associated with scientific men to carry on my work by myself, for I see that my mind is just as far in the regions of the unknown world searching every dark cavern for treasure of knowledge as theirs are."[38]

Chanute would later make the same kind of offer to others that he was now extending to John. The future recipients of his support would include

Augustus Herring, Edward C. Huffaker, Mouillard, William Paul Butusov, and William Avery. But his acknowledgments of John's accomplishments and potential were not enough to achieve the collaboration that he was hoping for. John shunned the overtures and remained on his own in Alton. The conference on aerial navigation had a far-reaching influence on the development of aeronautics in America, but even though it had brought recognition to John, he remained unwilling to share all that he had to offer.

CHAPTER 8

Other Pursuits

John recognized that true progress in the development of flying machines would require not only ample time and money but also an outlet through which theories could be tested empirically. In the late 1890s, both Mount St. Joseph's College and Santa Clara College were struggling financially and could offer nothing beyond a modest salary (roughly $30 per month) and laboratory space. Lacking an interest in business, John was left with no clear path to translate his aeronautical theories into practical form unless he was willing to forgo ownership of his intellectual property. During this crucial period in the development of aviation, he became increasingly isolated and unaware of advances being made by others, including Chanute and his assistants.

While some experimenters with heavier-than-air flying machines received recognition (and even some degree of acceptance) through publications such as Chanute's *Progress in Flying Machines* (1894), James Means's *Aeronautical Annuals* of 1895 through 1897, Matthias Forney's *Aeronautics* (the former *American Engineer and Railroad Journal*), *Scientific American,* and popular serials such as *Criterion Magazine, Century Magazine,* and *Cassier's Magazine,* the general public still viewed the prospect of human-controlled heavier-than-air flight as a dubious proposition at best. At the same time, and due in part to the interchange at the 1893 conference, certain networks of aeronautical researchers began sharing information, offering research and publication support, and sponsoring competitions related to aeronautics. The primary groups of aviation researchers in the United States were centered in New England (including James Means, Samuel Cabot, A. Lawrence Rotch, Albert A. Merrill, and Charles H. Lamson), Washington, D.C. (Alexander Graham Bell, Albert Zahm, Emile Berliner, and Samuel Langley and his assistants Edward Huffaker, Carl Barus, Charles Manley, and Augustus Herring), and Chicago (Chanute and his assistants Augustus Herring, William Avery, and Paul Butusov). Another important center for aeronautics was established by Carl Myers and his wife, Mary ("Carlotta"), through their "balloon farm" at Frankfort, New York

In early 1895, John forwarded his "Soaring Flight" manuscript to Chanute with the understanding that Chanute would use his connections to get it published. In the meantime, John turned his attention to teaching and to invention in more pragmatic fields. He improved the efficiency of hot-blast furnaces through an invention that he described as "a petroleum-burner attachment to the door of any furnace."[1] It utilized strategically placed air vents that could be regulated so as to produce simple or compound rotations of the flames within the burner, thereby aiding in the effective combustion of crude and heavy oils such as petroleum. Other improvements produced more thorough combustion, noise reduction, and relatively smoke-free exhaust. Because the primary end users for furnace-based heating during this time were in northern Europe and North America, John applied for international patents. In the fall of 1895, he received patents for his petroleum burner in the United States (Patent No. 549,679), Great Britain (Patent No. 21,477), Germany (Patent No. 88,977), and Canada (Patent No. 50,585). Whether he received any financial reward from this invention is unknown.

When Chanute received "Soaring Flight," he reviewed it, then forwarded it to Matthias Forney, the editor of the *American Engineer and Railroad Journal.* The 131-page manuscript, heavy on physics, was far too voluminous for most journals, and in the spring of 1895, Forney notified John that he would not be able to publish it. Chanute later told John that he disagreed with some of John's applications of Newtonian mechanics in his circulatory flow model of lift. Nonetheless, he did offer to submit it to the publishers of *Scientific American,* even though he doubted they would publish it.[2] John may have concluded that Chanute's criticism and his failure to help get both this new manuscript and his second Chicago lecture published were an intentional response to John's unwillingness to collaborate with Chanute the previous year. He expressed his frustration to Chanute in early August 1895, suggesting that his work had been intentionally suppressed. Although Chanute clearly did respect John's work, their differences regarding some aspects of John's theoretical approach remained unresolved in subsequent correspondence. They lost touch after the summer of 1895, and John then redirected his energies elsewhere. He set his aeronautical work aside and hoped that the future would bring other opportunities for financing the practical application of his flight technology.

During the academic break in the summer of 1897, John stayed with the family in Los Angeles and at Fruitland. While he was in Los Angeles, he conducted further experiments in fluid dynamics when time permitted.

He accepted an invitation to become a member of the Southern California Academy of Sciences and became more active in the organization. In September, Mount St. Joseph's College was closed after the death of its founder, Father Patrick Henneberry. John then left Alton and returned to Los Angeles to live with the family. During this period he also made several trips to Santa Clara to visit friends such as John and James Leonard and Father Robert E. Kenna, S.J., who as the vice president of Santa Clara College was now in line to become the school's president.

At the November 9 meeting of the Southern California Academy of Sciences, John presented some of his aerodynamics theories from 1893–95 for the first time in an invited lecture titled "The Mechanics Involved in a Bird's Wing in Soaring and Their Relation to Aeronautics." It was published in its entirety in the *Los Angeles Herald* on November 10, 1897, and received lengthy editorial comment the following day in the *Los Angeles Morning Herald* as well as the *Los Angeles Evening Express*. "The lecture was illustrated throughout with chart drawings," noted the *Express*, "and was listened to with studious interest by the audience."[3] This talk provoked significant local interest. Eleven years following Israel Lancaster's influential but controversial lecture in Buffalo, and at a time when Frederick Lanchester's theories were being suppressed by his peers in England, Montgomery received a decidedly more positive reaction in the western United States.

In the fall of 1898, John accepted a teaching position at Santa Clara College, where he served as professor of English and mathematics during his first year.[4] The addition of an inventive lay professor with a background in classics, physics, and mechanics helped to round out the teaching staff. Father Kenna became college president and saw to it that John was given a light teaching load in order to free him up for research and development in physics. Historians have expressed some confusion concerning John's professional title while working at Santa Clara College. The school catalog lists him variously as "Professor of English and Mathematics" (1898–99 academic year), as "Professor of Mathematics" (1899–1900 academic year), or in subsequent years as "Electrician."[5] John identified himself as "Professor of mechanics and assistant professor of Physics, Astronomy and Electricity," but he typically signed correspondence as "Assistant Professor of Physics."[6]

In the nurturing environment of Santa Clara College, John engaged in research and development and took part in near-daily religious activities, surrounded by friends and colleagues. He remained cautious, however, about experimenting with small-scale flying models, hesitant to reveal crucial details to well-educated students. Although the college provided a secure environment in which to work, he always kept his lab locked when he roamed the campus grounds. While he was at Santa Clara, John continued his friendship with fellow instructors John and James Leonard. He occasionally spent school holidays at their homestead, a large ranch called the Rancho San Andreas on a secluded stretch of the Monterey Bay coastline in Santa Cruz County. The creeks that met Manresa Beach, which fronted the southern edge of Rancho San Andreas, contained minerals (cassiterite, magnetite, and fine-grained gold—what the locals referred to as "black sand") within the sandbars that, because of their heavier densities, were mechanically sorted and concentrated by wave and current action.

Between 1899 and 1903, John developed at least two gold-sorting machines that utilized mechanical separation techniques and strong magnets to concentrate the heavier minerals, then separate the gold from the magnetite. These "gold concentrators" were likely first tested at the beach fronting Rancho San Andreas. John was awarded a U.S. patent (No. 679,155) on July 23, 1901, for his concentrator, and another (No. 742,889) on November 3, 1903, for a modification of the device. He also received a Canadian patent (No. 70,319) on February 19, 1901. He shared the technology with the Leonard family, who independently pursued its economic potential, eventually obtaining enough gold from Manresa Beach to finance the construction of a few small commercial buildings in the hamlet of Aptos, located about three miles northwest of their ranch. In the summer of 1900, Richard Montgomery traveled to Anchorage, Alaska, to visit James, who had moved to the Klondike region to prospect for gold. A secondary purpose of the visit was for Richard to evaluate the potential commercial market for John's gold concentrators, although his conclusions concerning the business potential of John's gold-sorting technology are unknown. They were specifically designed for the extraction of gold from placer deposits, and that was the primary mode of mining in the Klondike.

In April 1900, Andrew P. Hill, an alumnus of Santa Clara College who had become a distinguished photographer, initiated a campaign to purchase a large grove of old growth redwood trees for preservation in an area known locally as "Big Basin" in the central Santa Cruz Mountains. Due to Hill's diligent pursuit of supporters, Santa Clara College and Father Kenna played a critical role in a strong political effort to convince the state of California to purchase the land and set it aside as a park. Thirteen educators from the Bay Area were organized into a committee by Hill to help determine what actions could be taken to preserve the trees, and Father Kenna asked John

Fig. 14. This group of prominent educators met at Stanford University in May 1900 and instigated a movement to protect a massive ancient redwood grove in Santa Cruz County through the establishment of California's first state park. Stanford's president, David Starr Jordan, is second from left; John Montgomery, representing Santa Clara College, is second from right. Under the leadership of Andrew P. Hill and Santa Clara College, Big Basin State Park was established two years later. (Courtesy History San José Collection)

to serve as the college's representative. A club was organized (the Sempervirens Club) as a voice for this effort, and John then gave a series of public lectures to generate additional support. After a lengthy uphill battle to push an appropriations bill through the state legislature, in March 1902, 3,800 acres in the Big Basin passed into the hands of the people of California, becoming their first state park, and preserving one of the most magnificent tree groves in the world for future generations to enjoy.

On September 3, 1900, while rushing to catch a streetcar near his home, Zach Montgomery suffered a stroke. Within twenty-four hours, he was dead. Eulogies were published in many western U.S. newspapers, emphasizing his accomplishments and remarkable life and describing him as "a southern firebrand," "a picturesque character," and "a pioneer." His lifelong friend Captain James Connolly commented, "Few men had more cultured or refined literary tastes and it was in speaking of things literary that you got at the inmost heart of the man."[7] Zach's friend Oliver Miles noted that "He was not a man who was moved by impulse to undertake a serious work, or to prosecute it in a half-hearted or indifferent manner, only to abandon it when difficulties and obstacles came in the course of its accomplishment, but of his subject he was a hard and profound student, his deliberations ripening into convictions firm and lasting, and when his purpose was fixed he pursued it with an unswerving determination and dauntless courage that characterized his work in whatever he undertook."[8] This same description could as easily have been applied to Zach's eldest son, John.

John's many activities in physics, aeronautics, and protection of the environment were valued highly by the Santa Clara College administration. On occasion he was tapped for other duties. In 1900, for example, he erected the college's eight-inch equatorial telescope and completed the exacting procedure required to properly configure it.[9] In 1901, he developed a graduated rheostat to produce special lighting effects for the college theater's production of *The Passion Play*.[10] That same year, he provided expert testimony about electricity in a court case involving his friend John Leonard, who was by then an attorney in Santa Cruz. In recognition of his achievements in theoretical and practical physics, Santa Clara College awarded John an honorary doctorate in physics in June 1901.[11]

In the fall of 1902, Father Richard Bell transferred from St. Ignatius College to join the staff at Santa Clara College as head of the scientific department and primary instructor of physics. Bell was the former classmate at St. Ignatius who had been responsible for John's teaching opportunity at Mount St. Joseph's College in 1894. Although he was now the department head, he and John, working as equal collaborators in the winter of 1902–1903, combined their considerable talents and began a directed period of research and invention in Hertzian wave theory and its application to wireless telegraphy. They quickly produced a basic telegraphic system consisting of a transmitter (a "Tesla Oscillator," developed years previously by Nikola Tesla) as well as a transformer that stepped up the standard voltage of 110 volts to 20,000 volts. They eventually developed their own receiver, which was far more delicate than the conventional ones typically used for this work.[12] With this system, they successfully transmitted messages between San Francisco and Santa Clara College.[13] They also explored the sensitivity of vegetation to electromagnetic waves and thereby discovered that large trees could be used to boost telegraphic signals.[14] Together they gave a popular

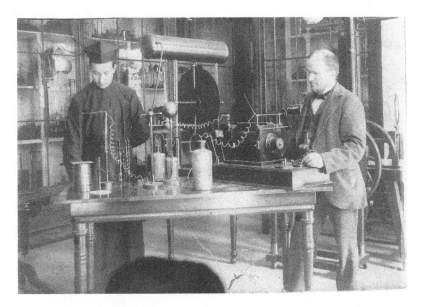

Fig. 15. Montgomery and Rev. Frederick Ruppert (substituting for Father Richard Bell) demonstrating wireless telegraphy as part of a lecture series on Hertzian wave theory at Santa Clara College, ca. 1901–1902. (Courtesy Santa Clara University Archives and Special Collections)

series of lectures and practical demonstrations of Hertzian wave theory at the college. Within a year of Father Bell's arrival at Santa Clara, their work in wireless telegraphy was attracting significant public attention.

Although John had made considerable progress in the areas of electricity, communication, and mining during 1895–1903, his momentum in the field of aviation had been all but completely lost. Others, however, had continued to make progress. By the turn of the century, the Wright Brothers had become interested in gliding. The owners of a small business engaged in the sale, repair, and manufacture of bicycles, Wilbur and Orville Wright began their shift to flying machines after studying the successes and failures of other aviators. The Wrights contacted Chanute for information, given that he was widely regarded as the most knowledgeable figure on the current state of heavier-than-air flying machines. Chanute later revealed that "When the Wrights wanted to start they wrote me and asked if I would permit them to use the plans of my biplane. They said they did not desire to enter aeronautics for financial reward, but as a pleasure or sport."[15] He added, "I turned over all my data to the Wrights on their application away back in 1900. I had gone as far as I could without a motor, and, as I was not mechanic enough for that and feared to injure the work by a failure in that direction, I left it to the Wrights, who were mechanics, to put on just the touches which were needed to produce success."[16] The Wrights soon began experimenting with biplane gliders patterned after those made by Chanute. They incorporated other elements obtained from the latest literature on flying machines. They had the financial means, the practical skills, and the benefits of Chanute's assistance in their advancement of aviation. As the proprietors of their own business, they also had the freedom and the money to pursue their new hobby whenever the demands for sales and repair work allowed.

While John had been busy with other pursuits, the Wright Brothers' relationship with Chanute had greatly accelerated their progress toward a practical and commercially viable powered flying machine. Through Chanute, the Wrights gained immediate access to Chanute's own experimentation with gliders, his prolific correspondence with like-minded aerophiles, his connections with influential members of the upper class, and his formal education and understanding of the U.S. Patent Office. In return, Chanute found in the Wright Brothers another means to help generate a practical solution to the flying problem on the strength of their skills in practical mechanics. The Wrights' work was a practical application of previous technology; their flying machine was a detailed progression of adoption, adaptation, and refinement. Chanute's "open access" policy regarding any and all information pertaining to flying machine technology allowed the Wright Brothers to leave their previous careers behind and enter the burgeoning field of aviation. Conversely, John's aversion to collaborate in aeronautics, the scarcity of aeronautical literature outside of major urban communities, and his location in Alton resulted in his further isolation. Increasingly, he lost the momentum gained prior to and including the Chicago conference, a trend that would continue over the next two years.

CHAPTER 9

A California Impetus

In an age when public exhibitions and aerial feats with hot-air balloons were popular with spectators, Thomas Scott Baldwin (1854–1923) began his remarkable career in the field of aerial exhibition. During his youth, he had worked as a tightrope walker in the W. W. Cole Circus in the 1870s, and he later performed trapeze work suspended from tethered balloons. In 1881 he relocated to San Francisco, where he worked in restaurants, occasionally supplementing his income at the amusement park at Seal Rock by performing tightrope routines between the famous Cliff House and a large rock just off shore.[1] Always on the lookout for new exhibition concepts, Baldwin heard of a well-known San Francisco–based balloonist named Park A. Van Tassel and sought him out.

Van Tassel had begun his career making balloon ascensions in New Mexico; he had relocated to San Francisco in the late 1880s. Inspired by the public exhibitions of Washington Donaldson, who was doing trapeze work from hot-air balloons, Park and his wife, Jeanette, made ascensions and performed a trapeze act while suspended from captive balloons for paying audiences at San Francisco's Central Park and the Seal Rock amusement park in the 1870s.[2] In late 1886, Van Tassel came up with the idea of making parachute jumps from ascending hot-air balloons as part of his act. He initially tested a 24-foot-diameter parachute by fastening a 150-pound sandbag to it and releasing the chute from a high point. Upon hearing of this work, Baldwin sought out Van Tassel, and together they became partners in a parachute jumping venture. However, after several weeks of instruction by Van Tassel and a few exhibitions, Baldwin suddenly abandoned the collaboration. The Van Tassels thought he had left the area.[3]

Baldwin had not disappeared entirely. Instead, he soon approached a promoter to propose a parachute performance of his own. He offered to jump from a balloon at an altitude of 1,000 feet, demanding a payment of $1 per foot of descent. The resulting public exhibition at San Francisco's Central Park on January 30, 1887, created an immediate sensation and was reported across the nation.[4] Moving quickly to capitalize on the stunt,

Baldwin embarked upon a national tour, performing at various venues. The novelty of Van Tassel's method was quickly diminished as a result, and when he arrived at a previously booked exhibition for a 4th of July carnival in Los Angeles, the carnival committee balked at the previously arranged fee. The Van Tassels desperately needed the money, so to ensure that they could collect on the promised payment, Jeanette appealed to the committee's sense of sensationalism by offering to make the jump herself. Described by one newspaper correspondent as "big [165 lbs.], young, handsome and blonde," she made her historic jump from an altitude of between 5,000 and 6,000 feet.[5] After her successful landing, she was immediately interviewed. "I ain't exactly a bird nor an angel," Jeanette remarked, "but it's just about what I imagine the sensation of flying is. It was beautiful!"[6] When asked why she had undertaken such a risky exploit, she defiantly said: "It is only a question of nerve. I made up my mind that I could jump from a balloon as well as Baldwin, and when I make up my mind to do a thing I do it. . . . I was anxious to get a reputation, and I did."[7] Thus began the career of the first female parachute jumper.

The Van Tassels and their unintentional protégé Baldwin popularized the sport and inspired a new generation of men and women aeronauts to parachute from ascending balloons for public exhibition. Following Jeanette's lead, several women achieved prominence in parachute jumping in the early 1890s, including Hazel Keyes of Sacramento, Leila Adair ("Lena Rayward") and Milie Anita of San Francisco, Louise Banner (a.k.a. "May Allison") of Cincinnati, and Katchen Paulus of Germany. Paulus co-invented the packed parachute system along with Paul Letterman. After the turn of the century, the enormously popular Georgia "Tiny" Broderick of Glendale, California, began her successful career in aviation making parachute jumps from balloons in 1908.[8]

Baldwin shared several traits with those who attained early prominence in aviation. He was a persuasive, audacious, and enterprising man with an uncanny ability to identify talent and capitalize on opportunities. With his deft exploitation of the Van Tassels, he soon eclipsed their influence in the sport and was able to gain considerable notoriety and financial reward. On his world tours in the late 1880s and 1890s, he enjoyed fame as the self-proclaimed "Professor" T. S. Baldwin.[9]

Baldwin's jumps launched the careers of a host of other California-based "aeronauts" who took up parachute jumping in the 1890s, including Bay Area residents Daniel J. Maloney and Frank Hamilton (who also took on the "obligatory" title of "Professor"). The most popular venues for these exhibitions were San Jose's Agricultural Park, Willow Park, and Alum Rock Park, Oakland's Idora Park, and San Francisco's Glen Park. With this sudden popularity, however, came an inevitable consequence: a significant increase in the number of injuries and deaths. Many people attempted their first jumps without any formal training whatsoever, and the increased pressure to add ever more elaborate stunts to the acts further decreased safety.[10] An aeronaut who stayed in the profession for any length of time was bound to eventually suffer some type of significant injury.

In the 1890s, Baldwin turned his focus to ballooning. He and his brother Samuel Yates Baldwin began manufacturing balloons at Quincy, Illinois. Baldwin became well-versed in the art of balloon science through his interaction with balloonists around the world.[11] Inspired by Alberto Santos-Dumont's successful and widely publicized public demonstrations with dirigibles in France, American contemporaries began to dabble in lighter-than-air flight technology after 1901. August Greth (San Francisco), Thomas C. Benbow (Montana), George E. Heaton (Oakland), George Tomlinson (New York), Carl Meyers (New York), and A. Leo Stevens (New York) were among the prominent Americans who contributed to the development of dirigibles.

In the fall of 1902, Baldwin teamed up with August Greth. Greth was a native of Alsace, France, who had become interested in balloons while serving in the French Army in Algeria, where balloons were used for reconnaissance. After studying medicine in Paris, he had moved to the United States, where he began to pursue dirigibles as a hobby in the early 1890s while simultaneously working as a doctor in San Francisco. He and some like-minded San Franciscans established the American Aerial Navigation Company (AANC) to fund the development of a dirigible.[12]

By late 1902, sufficient AANC stock had been sold, and Greth's plans to build a dirigible began to take shape.[13] Construction ultimately cost $12,000.[14] With his knowledge in ballooning, Baldwin was hired by the AANC to assist and to serve as the aeronaut. He oversaw the construction of the balloon (including the stitching of the silk panels, the proper mixing of gas, and inflation) but acknowledged: "I am a balloon man by profession and don't know much about motors or other machinery."[15] Through this symbiotic collaboration, however, Baldwin gained significant knowledge in dirigible mechanics and propulsion. Construction of the airship continued between October 1902 and January 1903 at the Rose Pavilion in San Jose's

Agricultural Park with the assistance of Mrs. De Witt Welch and Lincoln Woodford.[16] The dirigible was christened by the AANC as *The California Eagle*. At this same time, Baldwin became aware of John Montgomery's reputation as an expert in aeronautics and sought him out at Santa Clara College.[17] Said John later, "Baldwin offered as his excuse for seeking my acquaintance a common interest in aerial matters, and particularly a desire to learn at first hand so much of my theories in this connection as I might be willing to divulge. . . . I told him frankly what I expected to do and just how I expected to do it."[18] Baldwin quickly recognized John's potential. This meeting between the two men would have far-reaching repercussions for both of them, and for California's place in the history of American aeronautics.

Over several months in 1903, Baldwin persisted in learning more about John's theories and devices in aeronautics. As John recalled, "he was so satisfied with my ideas that if I would put them in a practical form he would have machines made and he would test them himself."[19] Baldwin came up with a business proposition: John's heavier-than-air flying machines would be exhibited in front of paying audiences by being launched at altitude from a balloon and then flown in their gliding descent by an aeronaut. Baldwin's proposed method for launching was a natural but novel extension of parachute releases from ascending balloons. This idea had been proposed by George Cayley in England, although it had not yet been attempted.[20] Through Baldwin's dogged persistence, and the promise of revenue for his research and development, John eventually warmed to the potential business arrangement.

In early June 1903, Baldwin wrote to John to report that Samuel Langley had successfully flown a heavier-than-air, unmanned, powered model flying machine called *The Aerodrome* in experiments conducted along the Potomac River near Quantico, Virginia.[21] The tests had gone so well that Langley was pursuing a larger, man-carrying version of his tandem-wing aircraft. The Langley design incorporated two wings positioned one behind the other in a tandem fashion rather than the more typical "biplane" arrangement common to the Chanute gliders. Upon reading about the Langley design, John was reminded of the similar tandem-wing arrangements he had considered back in 1883 as a result of observing pelicans flying in close formation.[22] Subsequent attempts by Langley on October 7 and 8 to launch a full-sized, piloted *Aerodrome* resulted in a well-publicized failure. The aircraft caught on the launching mechanism as it left the launch facility, which sent it plunging into the water below. *The Aerodrome* was known to be "a perfectly good flying machine" that was "excellently constructed."[23] However, the project had received significant funding from the U.S. Board of Ordnance and Fortification and the Smithsonian Institution, and the un-successful trials were sensationalized by the media as a waste of taxpayer funds. Although Alexander Graham Bell and Octave Chanute repeatedly defended Langley's design and attempts, *The Aerodrome*'s failure marked the end of both Langley's funding and his remarkable experimentation with heavier-than-air flying machines. Aviation pioneers Hiram Maxim and Clément Ader had also expended considerable sums on projects to achieve powered flight. These widely publicized efforts further illustrated

Fig. 16. John Montgomery, ca. 1903. (Courtesy Santa Clara University Archives and Special Collections)

to prospective experimenters that significant financial resources might be required for the development of a powered flying machine.

On October 19, Greth attempted his first untethered manned flight with *The California Eagle,* sailing over the city of San Francisco for two hours. He was unable to navigate back to his starting point at the AANC yard at the baseball grounds on Market Street because of the strong prevailing winds, but this flight is considered by many to be the first successful flight of a navigable dirigible in the United States. The newspapers immediately proclaimed *The California Eagle* as America's answer to Santos-Dumont's airship, and focused on the $100,000 prize awaiting the winner of the aeronautic competition at the forthcoming 1904 St. Louis World's Fair.[24] The announcement of the competition, and in particular the sizable prize, did not escape Baldwin's attention. While enjoying associations with both Greth and Montgomery, he discreetly developed plans for his own dirigible and traveled to Los Angeles to court potential financiers.

In the fall of 1903, John reentered the field of applied aeronautics. In late November he carried out a series of experiments to retest his hypotheses with small flying glider models before transitioning to man-carrying gliders. The model gliders combined his early control technology (1885–86) with aerodynamics theory (1894–95), and with his continued desire for stable flight, which he referred to as "equilibrium." Much as he had during his experimentation in the 1880s, John started with a basic airframe and then tested slight variations in airfoil configurations. "I simply used the warping idea which I had developed in 1885 and 1886," he said.[25] He brought three free-flight glider models to Rancho San Andreas in the fall of 1903: a 3½-foot-wingspan model with surfaces covered in pink silk called *The Pink Maiden,* a 5-foot-wingspan model with surfaces covered in calico fabric called *The Buzzard,* and a 7-foot-wingspan model that remained unnamed.[26] Each was designed to sustain various loads, even up to 7 pounds.[27] After reviewing these models with John, engineer and aviation author Victor Lougheed later provided a detailed description:

> In all of these, there was the application of certain definite principles. First was the use of proper wing curvatures to secure a maximum sustention with a minimum forward resistance, by a proper modification of the wing curvatures from center to tip; and second was the securing of automatic equilibrium by the principle of a large vanelike fin at the rear of the machine, so that any tendency to turn over, either by the impact of a wind gust or a sidewise inclination of the machine, must immediately result in its front or rear portion skidding sidewise, so that the tendency

to tip over would be terminated by a continuance of the glide in the direction of the fall.[28]

In John's three glider models, the location of the vertical "vane-like" fin was a central point of variation, and was an improvement upon his three earlier gliders from the 1880s.

Initial experiments with the models at Rancho San Andreas and at Santa Clara College showed great promise. Concurrent with these experiments, John attempted to learn as much as he could from Baldwin about the construction and inflation of balloons. In mid-November, he wrote to Baldwin in Los Angeles to report that the performance of the model gliders had confirmed several theories.[29] By that time, however, Baldwin had secured the financing he needed and had begun construction on his own airship. Despite his recent exposure to Greth's engineering expertise, Baldwin still lacked sufficient knowledge to produce an efficient propeller for the airship. Propellers available in the period were typically based on the marine propeller model. After initial pleasantries in his response to John, he broached the subject:

> Have you any special propeller that you think I could use on this airship? I am now at work on one that would carry out your idea and would not conflict in the least with your pet idea. If you have not, I could have one built and try it out. You know as well as I do that all of the airship people are at sea in regards to propellers.[30]

Without a written agreement, John was reluctant to divulge any thoughts he might have on propeller design.[31]

Only three days later, on December 17, 1903, the Wright Brothers made four straight manned, powered flights at Kitty Hawk, North Carolina, with their *Wright Flyer I*. Orville's logbook provided a firsthand account:

> I found the control of the front rudder quite difficult on account of its being balanced too near the center and thus had a tendency to turn its self when starting so that the rudder was too far on one side and then too far on the other. As a result the machine would rise suddenly to about 10 feet and then as suddenly, on turning the rudder, dart for the ground. A sudden dart when out about 100 feet from the end of the tracks ended the flight. Time about 12 seconds (not known exactly as watch was not promptly stopped).[32]

The second (piloted by Wilbur) and third (piloted by Orville) flights took similar pitching paths. In reference to Wilbur's first attempt, Orville said: "the course was about like mine, up and down but a little longer over the ground." On the fourth and final trial, with Wilbur at the controls, after traveling several hundred feet with occasional pitching, the machine finally darted to the ground, with the result that the front elevator was "badly broken up." Interestingly, when the brothers offered their first public statements to the press, which were published in the *Chicago Daily News* (January 6, 1904) and other newspapers, the story had changed. They offered a simpler description of the flights than the one in Orville's journal, emphasizing the presence of gusty winds, to which they now attributed the control problems, while dropping all reference to the crashes that had ended two of the four flight trials.[33] Subsequent published statements by the Wrights simplified the description even more, omitting any mention of control problems or even the presence of windy conditions. As a result, by the time of their first official public flights five years later, Americans were given the distinct impression that the trials of December 17, 1903, had been an overwhelming success in controlled and sustained flight. Nevertheless, although modest, these were the first successful powered flights made by man with a heavier-than-air flying machine. It would take nearly two more years and additional design variations (the *Wright Flyer II* and the *Wright Flyer III*) before the Wrights could demonstrate a practical powered aeroplane in the fall of 1905, and an additional four years to impress upon their advocates and the public the symbolic achievement of the initial powered flights of December 1903. Only a handful of reports of those initial flights appeared in newspapers in the western United States. John remained unaware of the Wright Brothers' accomplishment for some time.

John visited Rancho San Andreas that December to conduct additional tests with *The Pink Maiden, The Buzzard,* and the third unnamed model. They were hand-launched from the coastal cliffs and also from a nearby railroad trestle at Manresa Beach. John stretched a cable across a small valley and anchored it between two ridgetops. The gliders were ballasted with small sandbags, the "controls" were preset, and the launch was trigger-released from the cable at a height of 100 feet. The gliders were released in a variety of orientations ("in an erect position, on their backs, from the head, then from the tail and from a tip of the wing") so that it could be determined whether the designs could automatically "right" themselves.[34] Tests were also made to determine the range of possible controls and their effect on the glide path. "And in every instance," John wrote, "they would turn like a cat, then come sailing down beautifully, it making no difference whether the wind was blowing or not."[35] He later said, "I found them perfect in their equilibrium and would follow any direction I chose by giving

Fig. 17. John seated in the first mockup of his tandem-wing glider design at Santa Clara College. This craft, with bamboo longerons and rounded wing tips, was probably constructed in the winter of 1903–1904 specifically as a mock-up to test control mechanisms, but was never actually flown. (Courtesy Santa Clara University Archives and Special Collections)

them the proper warping."[36] On the basis of these experiments, John settled on *The Pink Maiden* design as the best for demonstrating control and equilibrium at full scale. This 1903 tandem-wing design remained his basic aeroplane design for the next several years. These results from Rancho San Andreas instilled a growing confidence in the conservative fathers of Santa Clara regarding John's eccentric passion for flight. In a letter written at the time, he told Ellen, "the Fathers who know about it are very enthusiastic."[37] On the strength of the tests with tandem-wing models, John constructed a new full-sized glider as a mockup with a new cruciform tail unit and a foot bar to provide wing-warping control.

In January 1904, John attempted to formalize his working relationship with Baldwin by sending him a written proposal. It was finally signed by both parties on April 28.[38] The contract required Baldwin to furnish the necessary capital (proposed at $1,000) for the construction of several gliders. John would supply the technology and instruction on methods of construction, and after they had achieved success with a full-sized version, Baldwin would provide the balloon, hire and train an aeronaut, and cover the expenses incurred in the course of exhibiting the glider by way of balloon-tethered launches at altitude. The glider was to be called the

Montgomery Airship. John would retain the patent rights, and Baldwin would retain half of any proceeds earned at exhibitions around the country over a period of two years.[39]

The contract also described John as "an inventor of a certain flying machine, device or contrivance for gliding and soaring through the air without the aid of gas, for the issuance of letters patent."[40] Thus, by April 1904, it was clear that John planned to secure patent protection for his tandem-wing glider design once its competency was proven on a large scale through fully controlled manned flights. Eleven years had passed since his participation at the 1893 World's Fair at Chicago. The aeronautical exhibitions at the forthcoming 1904 World's Fair in St. Louis would give him an opportunity to demonstrate his invention in front of his peers. Thus the contract stipulated that the glider was to be exhibited at St. Louis specifically for "scientific and educational purposes" rather than for prize money.[41]

With their business arrangement formalized, John committed himself to the project. After roughly twenty years of discreet and independent development in aerodynamics and aeronautics, he finally found himself with the finances, employment, and practical means (both material and time) to apply all of his theories and engineering to developing a machine that would be capable of stable, controlled flight. However, when it came to the *Montgomery Airship* project and the aeronautic competition at St. Louis, Baldwin had something entirely different in mind.

CHAPTER 10

A Rude Awakening

It was Professor J. J. Montgomery who brought to light again and developed the old well-known principle of the parabolic curve, which is one of the vital points in the construction of a propeller of an airship.

Thomas S. Baldwin, 1908

Between mid-April and mid-May 1904, at the same time that Baldwin was helping to construct Montgomery's full-sized tandem-wing glider, he was occasionally assisting August Greth with test flights of *The California Eagle* at San Francisco. On April 23, the dirigible, piloted by Baldwin, was flown at the baseball grounds at 11th and Market Streets. After it had ascended a few thousand feet and made some initial maneuvers, the change in altitude required an adjustment of the carburetor. Knowing little about these mechanical subtleties, Baldwin could not prevent the engine from cutting out altogether. He fumbled ineffectively with the engine while Greth and his stockholders watched their $12,000 investment drift aimlessly southward over the city. Baldwin and *The California Eagle* finally landed in a grassy area along San Bruno Road in South San Francisco, resulting in damage to the gondola's aluminum frame. In an interview immediately following the incident, Greth quipped, "He may have been capable of handling the steering apparatus with which I have equipped the machine, but the motor evidently bothered him."[1] Not only had Baldwin placed himself and the dirigible directly in harm's way, but he had generated an unnecessary public failure for Greth. As a result, Greth immediately disassociated himself from Baldwin.

Montgomery and Baldwin, however, continued their mutual project. They worked together on experiments for several weeks at Santa Clara College in April and May. Out of curiosity, Father Bell occasionally observed some of these experiments. With the reassurance of a formal business partnership, John revealed many of his discoveries about the design and construction of and the theory behind his heavier-than-air machines. Baldwin eventually raised the subject of Montgomery's propeller design, and that led John to demonstrate his theories through a working propeller model, which was subjected to tests in a wind tunnel he had designed at the college. Baldwin's descriptions are the only surviving accounts of these experiments. He wrote:

> We experimented for weeks in the placing of the blades on the shaft, that they might work on a column of air equal to the periphery of the propeller. We found to do this we must bend the blades back 5 degrees. At less than this angle the air rushing around them buckled back and set up a current of resistance in again coming in contact with the blades. At a greater angle the air slipped from the blades without their attaining a full clutch upon it. We figured out the proper velocity of our propeller. It must go forty-seven miles an hour in its circular traveling. This is 200 revolutions a minute. To go beyond that was to lose power. In the solid aluminum propeller, curved true to an eighth of an inch, and with smooth, metallic surface, we are near perfection in this important feature of the driving mechanism of the airship.[2]

In a separate article, he provided further detail:

> At first we had five or six screws [blades] on the propeller, but we finally got it down to only two blades, and the curve of those represented the most important point in the whole proposition. To have this parabolic curve so as to have each screw absolutely clear and utilize the "air-wake," or air disturbed, at the same time avoiding all possible resistance, was the important question we had to consider. We found by a series of experiments that the air in striking a blade of a parabolic curve at a certain angle—say forty five degrees—goes up over the blade instead of down as commonly believed. . . . In the construction of the [*California*] *Arrow* we utilized all the information obtained by these experiments, and also were careful about selecting materials.[3]

John's propeller design was a natural extension of his airfoil experiments in the 1880s and early to mid-1890s. Through this instruction, Baldwin learned about proper twist, pitch, and other aspects of airfoil/propeller blade design and construction that were significant advances over the more standard marine and windmill model approach.

In May, construction began on a full-sized version of *The Pink Maiden* at Santa Clara College. The resulting unnamed Montgomery-Baldwin tandem-wing glider weighed 32 pounds and had two identical wings set one in front of the other, spanning 20 feet in length by 3 feet wide and providing a total of 120 square feet of lifting area. Toward the end of the month, John sent the machine to Peter A. Cox's Pina Carta Ranch in the Gabilan Mountains just south of San Juan Bautista, California, with the intention of conducting trial flights. On the evening of their planned departure, Baldwin abruptly announced that he was obligated to go to Los Angeles on business, but he promised to return within a few days and join John at the Cox homestead to help with experiments.[4] The glider was disassembled and shipped in a pine crate by train to the Leonards' Rancho San Andreas, then by wagon to Cox's ranch. Once it was safely there, John mailed Baldwin directions to the ranch in anticipation of his arrival. Despite his promise to assist, however, Baldwin never showed, and John was left to finalize the construction and test the glider himself.[5] He finished assembling the glider and constructed a derrick launching apparatus on top of a steep hillside consisting of a cable stretched between two high poles. With John at the controls, the craft was raised and released, after which it glided down the mountainside, "meeting [the] most erratic gusts of wind."[6] He "made a number of successful flights" in June with the occasional assistance of Peter Cox and his ranch hands.[7]

In addition to the derrick launching method, John launched some of the flights by simply running downhill while holding on to the frame. During these glides, he observed how rapidly the large vertical fin responded to side gusts. As he later wrote, "I was gliding down the hill when quick as a flash I was whirled at right angles to the first wind, but was not upset."[8] The glider was also flown as a kite with a tethered control bar dangling from its center of gravity, which John could use to test wing-warping control while the machine was in flight. In August, the glider was transported back to Rancho San Andreas for further evaluation. Hoisted by a heavy cable, it was suspended across a canyon at the ranch, in an arrangement similar to prior experiments. Using a trigger release mechanism and fixed preset flight controls, the glider was ballasted with sandbags to simulate the roughly 150-pound weight of a pilot. It was launched from a variety of orientations (from the wing tip, upside down, from the tail, etc.) so that its flight characteristics could be observed. In one experiment, wooden rails were attached to the bottom of the frame, and the ballasted craft was launched from the 100-foot-high coastal bluff toward the beach. An extant photo shows that this makeshift sled at the base of the glider left long grooves in the sand after an unmanned glide. The aircraft had a very shallow gliding angle, and the tests were successful.

John remained perplexed by Baldwin's failure to return from Los

Fig. 18. The full-scale 32-pound glider of the Montgomery/Baldwin project, ballasted and released (unmanned) from the nearby coastal cliff at Manresa Beach, in August 1904. Note the grooves left in the sand by the frame, indicating a very shallow angle of landing. (Courtesy Santa Clara University Archives and Special Collections)

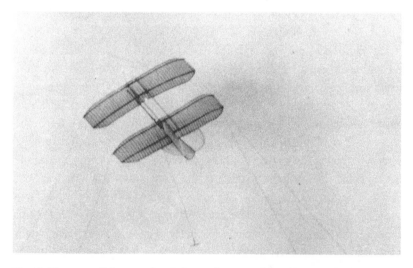

Fig. 19. The same glider as in fig. 18, being flown as a kite at Rancho San Andreas in August 1904. Note the tethered control bar dangling underneath for operation of the wing-warping system. (Courtesy Santa Clara University Archives and Special Collections)

Angeles. After two weeks and three letters from John, Baldwin finally responded, only to say that he had been "given a chance to build a small airship while in Los Angeles and had shipped it to San Jose where he was completing its construction."[9] In July, he contracted bicycle repairman/designer Charles V. Randall and dressmaker Mrs. De Witt Welch to perform much of the construction and fabric stitching. The dirigible was built discreetly in the yard behind Randall's Cyclery.[10] After making some inquiries, John found Baldwin at the Cyclery. They talked for a while, and then Baldwin led John into the yard in back of the shop and showed him a balloon airship in the process of construction. John realized that Baldwin had been utilizing principles obtained through their confidential interchange of ideas at Santa Clara College during the previous months: "The propeller in particular was an adaptation of my experiment."[11] Baldwin turned to John and commented, "Well Professor, made right?" John simply departed in silence.[12] The unpleasant reality of their "collaboration" was that Baldwin had manipulated John in order to extract knowledge from him. However, Baldwin now found himself in a predicament: after boldly appropriating technology and abandoning their written contract, he had no idea what John might do next. With his investors expecting him to take the $100,000 prize at St. Louis with his new dirigible, he attempted some damage control. Two weeks later, John noted, "I received a letter from Baldwin in which he said the airship I had seen at Randall's yard was a flat failure, and that he was going to store it." History shows however, that shortly after their confrontation at the Cyclery, Baldwin moved the dirigible to Oakland's Idora Park, and within the very same two weeks, he completed its construction.

While he was working on the airship, Baldwin noticed a young man riding a motorcycle nearby. He was impressed with its power for its relatively small size. The motor was a 2-cylinder, 7-horsepower engine, constructed by the Glenn H. Curtiss Manufacturing Company of Hammondsport, New York. Baldwin immediately wired the company and placed an order for a Curtiss engine, hoping to adapt it for use on his dirigible. It would be another in a series of fortuitous decisions by Baldwin during the period 1903–1904 that would forever change American aeronautics.

The California Arrow consisted of a balloon made of oiled Japanese silk with a capacity of 8,000 cubic feet and a 30-foot-long frame using the Curtiss 2-cylinder, 7-horsepower engine.[13] A feature article in *Scientific American* described it as "having two metallic blades, and nearly six feet in diameter."[14] On August 3, 1904, Baldwin ascended in the airship after taking off from Idora Park, traveled ten blocks in a slight breeze, turned around,

Fig. 20. Thomas S. Baldwin proudly displaying his propeller from the airship *The California Arrow.* The (barely visible) flattened wood ribs sewn within the canvas on the propeller helped to maintain its twist while also being as lightweight as possible. At St. Louis he had used a heavier, metallic blade. (From Elbert E. Dewey, "An Airship's Success," *The Technical World* 2, no. 1 [September 1904]: 476–84)

and returned to his starting spot—a successful round-trip flight. *The California Arrow* has thus been recognized as "the first dirigible in the United States to make a start, make a trip and return to the starting point without incident or accident."[15] Baldwin's melding of Greth's mechanical knowledge, Montgomery's propeller design, and Curtiss's motorcycle engine was an unparalleled advancement in American lighter-than-air technology and performance, all with Baldwin perceived as the sole inventor.

At roughly this same time, successful experiments with a tandem-wing glider were being conducted by Louis Paulhan and Louis Peyret (Louis Blériot's foreman at his aviation works) at Captain Ferdinand Ferber's aerodrome near the Promenade des Anglais in Nice, France. Tandem-wing designs had been conceived of by Thomas Walker (1831), a Frenchman named Danjard (1871), a man named Prigent (1871), Samuel Langley (1892), and Lawrence Hargrave (1893) and had been flown by Wilhelm Kress (based on a design from 1877) in 1901. Paulhan and Peyret's glider had wings set at a dihedral angle, 225 square feet of surface area, an operable elevator as a canard, and a total weight of 90 pounds.[16] Peyret became a strong believer in the potential of the tandem-wing arrangement, and went on to design and build a successful tandem-wing powered craft for Blériot in 1907 (the *Blériot VI Libellule* [Dragonfly]), in which flights up to 490 feet were made. In September 1904, at Welsh Harp in the United Kingdom, a Brazilian referred to in the literature as Señor Alvarez made a high-altitude unmanned, ballasted launch via timed fuse of a powered aeroplane from an altitude of approximately 3,000 feet.[17] The craft was equipped with a 2-horsepower Minerva engine powering two counter-rotating propellers. Lateral stability was provided by wings set at a dihedral angle.[18] The cruciform tail included a stationary elevator and a movable fin. Upon release, the machine initially took a dive, but it then leveled off for a flight of about one mile.

In late October 1904, Baldwin shipped *The California Arrow* to St. Louis in time for the World's Fair. There he met a balloon enthusiast named Augustus Roy Knabenshue, a native of Toledo, Ohio, who was giving rides to fairgoers on a tethered hot-air balloon. Knabenshue agreed to pilot Baldwin's *Arrow,* and after some instruction he quickly proved to be an adept pilot. He subsequently made several successful flights, most notably on October 25, when he piloted the dirigible for eleven miles to secure first place in the aeronautic competition. *The California Arrow* was proclaimed the most successful airship to date.[19] After a heated discussion with the aeronautic officials over compensation, Baldwin packed up *The Arrow* and left the World's Fair grounds for a promotional tour of the Northeast, including Curtiss's facility at Hammondsport, New York.[20] On November 21, he applied for a U.S. patent for "a new and improved airship" that covered several improvements, but curiously not the rudder or propeller.[21]

John learned of Baldwin's accomplishment through a newspaper article describing *The California Arrow*'s success at St. Louis. He immediately recognized *The Arrow* as the same dirigible he had observed at the Cyclery in July—the so-called failure that had supposedly been put in storage.[22] He now knew that he had been lied to, but with Baldwin constantly on the move, it was nearly impossible to pursue him for breaching their contract.

Although Baldwin and Knabenshue returned to Los Angeles as part of their national exhibition tour, they specifically avoided exhibiting *The California Arrow* in the Bay Area. *The Arrow* received widespread exposure without proper credit being given to Greth or Montgomery, and with only tangential credit to Curtiss for his motorcycle engine. Baldwin pocketed a

sizable income from the tour and enjoyed new prestige. He had a keen ability to translate others' ideas into effective platforms and then capture the
imaginations of both investors and the paying public. Within the space of
roughly twelve months, the man formerly described by others as the "Father of the Parachute" and self-described as "merely a Balloonman" had
been promoted to America's "Father of the Dirigible," marking the start of a
career in which he would prosper for the rest of his life.

In a feature article on *The California Arrow* in the *New York World*
in late November 1904, Baldwin acknowledged that he had studied at Santa
Clara College for an extended period "under two eminent teachers in physics, Father Bell and Prof. Montgomery," and implied that Baldwin had been
an equal collaborator in experimenting with and perfecting a successful
propeller design.[23] After finally receiving several letters from John, Baldwin
returned to California and went to see him. During this stormy meeting,
John confronted him on his bold appropriation, as well as the more serious financial unfairness. Baldwin laughed, but admitted that he had made
use of principles received in confidence. John later recalled, "when I appealed to his sense of fairness he told me that 'money' was his honor, and
we parted without any settlement or accounting."[24]

Shortly after this encounter, Baldwin was interviewed by the *San Jose
Mercury.* He again emphasized the scientific training he had received at
Santa Clara College, but this time he did not even mention Montgomery,
implying instead that Father Bell was his instructor.[25] This was the final
straw for both John and Father Bell. Bell countered Baldwin in an interview published in the *Mercury* the very next day. "Santa Clara College, the
pioneer institution of learning and scientific knowledge on this coast," he
wrote, "has had some very able men attached to it in every branch of learning from the very beginning, and I believe it has not yet departed from its
glorious traditions. Professor J. J. Montgomery is a man of whom Santa
Clara College is justly proud, and it is absolutely to him alone that Capt.
Baldwin is indebted for the success attained in the management (control
and guidance) of his airship at the World's Fair."[26]

Despite this public challenge, Baldwin and Knabenshue returned to Los
Angeles, where they based further exhibitions out of Chutes Park, including a sensational race between a Pope-Toledo automobile (driven by Bar-
ney Oldfield) and *The California Arrow* (piloted by Knabenshue). These
events stimulated public interest in the dirigible, all the while exposing the
successful airship technology to a community of aerial enthusiasts and in-
spiring a wave of invention in lighter-than-air machines in Los Angeles,

including *The Bullet* by L. B. Haddock and Wordin Trombley in March 1905, and Alva L. Reynolds's *The Sky Cycle*. Chutes Park became a hub of activity, with the spontaneous invention of various airships, some of which were identical to *The Arrow*.[27] By August 1905, this activity led to the formation of the Los Angeles–based Southern California Aerial Navigation Club (SCANC), with Wordin Trombley serving as president.[28] Trombley and Reynolds then participated in a SCANC-sponsored airship competition at Chutes Park on September 10.[29]

In the winter of 1904–1905, Baldwin visited Glenn Curtiss in Hammondsport to discuss a future collaboration on propulsion systems for Baldwin's airships. Baldwin's audacity and the fact that he had repeatedly exposed the technology he had appropriated fueled John's ambition to establish his priority sooner rather than later. In essence, he would have to fulfill his end of the Montgomery–Baldwin contract. In November and December 1904, John launched into a new phase of gliding experiments in seclusion at Rancho San Andreas. Rumors circulated quickly to neighboring communities. In order to ensure secrecy, the Leonards told the public that John was experimenting with a new type of parachute.[30]

While poised to give others a working demonstration of his flying machine for controlled flight, John had completely lost all means to accomplish this without the assistance that Baldwin was to have provided. He was left with a decision to make: How could he best continue progress, in an adversarial environment, with little capital at his disposal? This challenge inspired Montgomery to fulfill his part of the agreement in a very public way.

CHAPTER 11

Going Public

That sudden thrill of almost horror, something entirely new to me, came just at that moment as I had realized that the balloon had gone and that literally I had to fly for my life. But all I had to do was grasp firm and glide the machine.

Daniel Maloney, 1905

Determined to prove the validity of the Montgomery aeroplane, Montgomery and John Leonard set out to acquire the means and manpower for flight experiments. Leonard called upon his friend Fred Swanton, a tireless promoter of Santa Cruz tourism who at the time was the manager of the Santa Cruz Beach Company. Swanton came to the Leonards' ranch for demonstrations of the model glider *The Pink Maiden* and witnessed its launch from the railroad trestle at Manresa Beach. Suitably impressed, he discussed the possibility of future exhibitions at Santa Cruz, including hiring a hot-air balloon, balloonist, and pilot. He suggested the idea of launching the glider from a captive balloon tethered 1,000 feet above the beach at Santa Cruz.

John subsequently visited the Montgomery family home in Oakland with *The Pink Maiden.* In a particularly good mood, he wrapped his arms around his mother and asked his oft-repeated question: "How's the boss?"[1] He gave the model to Ellen as a gesture of affection, saying that he always wanted that particular model to be associated with her. He gave her an affectionate peck on the cheek and pointed out to her that he had sewn the fabric on the wing himself, adding, "You didn't know your son could sew on a machine, did you?"[2]

Launching manned gliders from high altitudes would naturally require a trained aerial specialist accustomed to such heights. Through Swanton's social network, John Leonard was introduced to Frank Hamilton and Daniel Maloney. Hamilton, who went by the show name "Professor Hamilton," hailed from Toronto, Canada, and was an experienced hot-air balloonist.

Fig. 21. Ellen Montgomery, ca. 1917, holding the tattered model of *The Pink Maiden*, which had been damaged during the 1906 earthquake. (Courtesy Santa Clara University Archives and Special Collections)

He performed parachute jumps and trapeze stunts from tethered balloons at county fairs. Maloney was a twenty-six-year-old native of the Mission District in San Francisco who since 1898 had used the moniker "Professor Lascelles" or "Jerome Lesalles" for his exhibitions of parachute jumps and trapeze work suspended from balloons.

Maloney was well-known throughout the San Francisco Bay area for his exhibitions at Glen Park in San Francisco and Idora Park in Oakland. He was accustomed to descending by parachute from altitudes of 500 to 800 feet, the minimum height considered practical for parachute inflation. Like so many of his colleagues involved in the field, he had been seriously injured more than once during his exhibitions. Despite the risks involved

Fig. 22. Daniel J. Maloney (ca. 1900–1905). Medals were customarily worn by balloonists and aeronauts at the turn of the century. (Courtesy Carroll F. Gray Aeronautical Collection, privately maintained)

in this line of work, by December 1904 he had already quit his job at a towel factory to become a full-time aerial exhibitionist.

John and the Leonards rented two hot-air balloons from Swanton, hired Hamilton to oversee their inflation, and hired Maloney as the aeronaut. Maloney was ideal for the job, as he possessed the nerve for high-altitude flight and a cautious, conscientious attitude. In order to avoid confusion about his own identity as a genuine professor, John insisted that both Hamilton and Maloney drop their assumed titles. Given that Maloney, at 145 pounds, weighed slightly more than John, a second glider was constructed at Santa Clara between January and February 1905 with a slightly larger wing area than the 32-pound glider of May 1904. This new glider weighed 42 pounds and had a total surface area of 180 square feet. It was shipped to Rancho San Andreas by train in early February. Maloney then joined the group, and over the next three weeks he received instruction on the aircraft's form and function, including demonstrations of unmanned, properly ballasted glider flights launched (by trigger release) at an altitude of 100 feet from a suspended cable, and also from the 100-foot-high railroad trestle at La Selva Beach. Much as he had done with the small-scale models, John preset the controls and launched the craft from a variety of positions. When released, the glider would right itself and continue on its course to a gentle landing at a very low gliding angle.

Hamilton joined the group near the end of February. A series of powerful storms hit California at just this same time, culminating on March 12 and 13 with widespread damage along the state's central coast, bringing all experimentation to a halt. With the uncertain but imminent timing of the glider exhibition, the local press followed the story. The *San Jose Evening News* mentioned the anticipated flights on March 8, 11, and 18, 1905.[3]

Also during the week of March 5, yet another successful American airship, George Heaton's *The California Messenger,* was made public at an exhibition at Oakland's Idora Park. Designed and constructed the previous fall by Heaton and his wife, Georgia, the dirigible was 76 feet long and 24 feet in diameter, with a capacity of 10,000 cubic feet of gas. After initial tethered trials in December 1904 and early 1905, Heaton employed two young aeronauts, Lincoln Beachey and David F. Wilkie, both of whom were active in the Bay Area's motorcycle racing circuit and the Pacific Motor Cycle Club. In his teenage years, Beachey had served as Baldwin's mechanical assistant at his San Francisco balloon yard. Both Beachey and Wilkie had remained in aeronautics: the former became an adroit dirigible pilot and only six years later was regarded as America's premier stunt flyer; the latter figured in Montgomery's career.

By March 16, 1905, the weather had improved sufficiently to begin balloon-launched flight trials with Montgomery's tandem-wing glider. Out of caution, John placed clamps on the controls to restrict their range of motion for the first trial, so that Maloney could only glide gently to the ground in shallow turns. He specifically instructed Maloney not to cut loose at altitude if anything appeared to be wrong with the glider, or if he had any doubts about making a flight. Should he choose to stay with the glider still hooked to the balloon, it would gradually descend as the balloon cooled, eventually affording him a gentle landing. This instruction remained standard operating procedure for all subsequent flights.

On the morning of March 16, after a failed first attempt to launch the balloon, a second attempt was successful. The hot-air balloon carried Maloney to a height of 800 feet, at which time the glider was cut loose; it descended in circles until it was within a short distance of the ground. At this point, and in accordance with John's instructions, Maloney flared the glider almost to a complete standstill using the elevator and wing-warping system. He landed in an apple orchard without mishap.[4] For the next trial, on the morning of March 17, John increased the range of motion on the controls and instructed Maloney to use the warping to control the glider through a series of prescribed maneuvers. The first attempt was abandoned because of difficulties with the balloon and glider. A second attempt was made that same day (the fourth trial in the series overall), resulting in further success. John wrote, "At an altitude of about 2,000 feet, the balloon and craft suddenly struck a strong wind which carried it rapidly away." When Maloney finally released at an estimated altitude of 3,000 feet, he "glided in spirals as directed, some very rapid and short, at these times the machine being inclined sideways more than 45 degrees."[5] On March 20, a third set of experiments was conducted, this time with Father Bell in attendance. John relaxed the control system even further in the hope that Maloney might make an even more extended flight. John recalled, "I cautioned him not to be carried away by enthusiasm at the sensation of floating at his own will through the air. . . . I especially warned the aeronaut not to attempt too much on this occasion. Following my instructions, Mr. Maloney cut loose at a height of 3,000 feet and glided to the earth in long sustained circles and with a speed which varied."[6] Maloney commented afterward, "For a few minutes I simply poised in the air and then flew around in different directions, circling, darting back and forth, up and down, as easily as an eagle could have done it. I was up in the air eighteen minutes and never had the least difficulty in guiding."[7]

Maloney's ability to choose his maneuvers and course, either with the wind or against it, at such great heights above the earth, and to land lightly on his feet at a predetermined location, was an unprecedented accomplishment in heavier-than-air flight of any kind. The flying machine worked precisely as envisioned, including three-axis control with maintained "equilibrium." In addition, the Aptos flights of March 17 and 20, 1905, far surpassed all previous records for altitude, distance, and duration for gliding flights anywhere in the world.

Accounts of Montgomery's success received widespread attention in American newspapers, with more muted interest in Europe. One of those reports caught the attention of Amos Ives Root, a primary advisor to the Wright Brothers. Root was a wealthy inventor and manufacturer of beekeeping equipment in Medina, Ohio, who had long had an interest in aviation, publishing occasional articles on the subject in his magazine *Gleanings in Bee Culture*. He had begun a correspondence with the Wright Brothers in early 1904 and soon became part of their social information network, providing information on other aviators, particularly whenever success was noted. Upon learning about Maloney's flights, Root immediately wrote to Wilbur Wright on March 22 and offered to make an appeal to some of his Santa Cruz County subscribers to hunt down some intelligence on the events at Aptos.[8] As the Wrights became aware of these successful controlled flights on the West Coast, and of Montgomery's interest in patent protection, they began actively tracking his progress.

As a gesture of gratitude to the fathers of Santa Clara College for their continued moral support, on March 20 Montgomery christened his successful 42-pound tandem-wing glider *The Santa Clara*. After learning of his success, Father Kenna, John H. Leonard, and others persuaded John to accept the public recognition he deserved. A press conference was arranged for the evening of March 25, so that Montgomery and Father Bell could provide firsthand accounts of the Aptos flights, as well as background on John's heretofore unpublicized activities in aeronautics. This public disclosure not only was a matter of due diligence by Montgomery but also served to cement his position as an independent originator of aeronautics despite Baldwin's recent misrepresentations to the public.

Although John had previously discussed his flying machine experiments of the 1880s at the Chicago conference of 1893, this pivotal press conference gave him the opportunity to publicize the fact that he had flown in the 1880s and that his methods for lateral balance and longitudinal control had been developed at Otay during 1884–86 with "more or less success."[9] He also noted that his aerodynamics experiments of 1894–95 had been instrumental in his solving the dilemma posed by his gliders of the 1880s using airfoil theory. "I started to build the models something over a year past, but I found it necessary to refer to the old formula and data and these were followed without deviation."[10] Despite these admissions of previous success, the press focused on the details of the Aptos flights and the precedent of controlled high-altitude flight.

John acknowledged that many problems with manned flight remained. However, he clearly believed that the Aptos flights had proved that he had successfully addressed the first and most crucial step in solving the "problem of aerial navigation." On the question of motive power, he said, "The question of motor can only be considered after the machines, as to their soaring powers are complete."[11] Picking up on this theme, an editorial in the *Washington Post* placed John's priorities in their appropriate context:

> The layman would reserve these problems and tackle first the one of rising from the ground. But the real aeronaut, like Prof. Langley and Prof. Montgomery, knows that this is a mere matter of detail, to be solved by the "application of proper principles." Continuance of flight, too, is a secondary matter of a man who knows his business. The great problem is guidance. . . . Naturally, as Professor Montgomery suggests, the size of the motors will be left undetermined until the machines, as to their soaring powers are complete. . . . The problem of aerial travel by man has been solved again—in part. There is danger that someone with willful ignorance of the immutable laws discovered and illustrated by Prof. Langley may fly away with the prize.[12]

The press conference represented a pivotal moment in John's career. However, he remained circumspect about specific details so that his patent rights remained protected.[13] He had already prepared specifications and drawings for his patent application on the control system for the aeroplane, so he refrained from displaying the glider in public.

On March 27, a story about Montgomery in the *Chicago Record-Herald* caught the attention of Octave Chanute. Chanute had also received press clippings on Maloney's flights from friends in Great Britain. Unaware that the Wright Brothers were already tracking Montgomery's progress, he notified them in a letter of April 4, enclosing "an account of a bold performance in California. I will write Montgomery for particulars."[14] On that same day, Chanute wrote to John:

I note from newspaper accounts that you have achieved great success in some flying machine experiments. I shall be glad to receive such published accounts as may be available, as well as a sketch or photograph of your machine when you are willing to communicate them. The accounts here are brief but seem to indicate that the machine is under perfect control.[15]

The following day, Chanute sent a letter to a colleague, Major Hermann W. L. Moedebeck. "I send you a newspaper report of a bold performance in California," he wrote. "As I know Montgomery personally I have written to him to ascertain what truth there is in this report."[16] Moedebeck was a prominent expert on European aeronautics, and was busy at the time preparing a revised edition of his *Pocket-Book of Aeronautics* in Germany for distribution in Europe and North America. Acknowledging America's leadership in aeronautics, Chanute added, "If it turns out to be correct, this together with the Wright performances [of 1904] will greatly increase the demand of your pocket book."[17]

In a lengthy response to Chanute, Montgomery revealed additional details of the craft and its performance in flight. Upon receipt of John's letter on April 11, Chanute forwarded it on April 16 to the Wrights with the instruction: "Enclosed please find a letter from Mr. Montgomery which please return when you have read it."[18] Amos Root also soon received additional correspondence from his California-based journal subscribers and newspaper articles on Montgomery and forwarded them to the Wright Brothers.

Newspapers in the western U.S. often connected Montgomery's success with Langley, Lilienthal, Maxim, and in some cases Mouillard or Le Bris, but not the Wright Brothers.[19] For the Wrights (and for Root), the general lack of recognition for their accomplishments was discouraging. Root repeatedly advised the Wrights to make their work public so that they could receive greater recognition. With Wilbur's editorial input, Root occasionally used his magazine as a forum to keep his readership updated on their progress. However, he expressed some frustration in a letter to Wilbur: "I for one am disgusted with the press generally, especially the scientific press, to find so few of them have taken the trouble to copy what I have written or make any mention of the matter."[20]

Whether out of envy or competitiveness, the Wrights orchestrated a public relations response within a week of receiving reports of Maloney's flights at Aptos. Prompted by a letter from Wilbur Wright, Root published an editorial in *Gleanings in Bee Culture* on April 1 titled "Wright Brothers Still Ahead." In it he stated his opinion that Montgomery's accomplishments were negligible when compared to the Wrights' achievements in flight to date, criticizing John's use of a balloon to hoist the glider to altitude.[21] After additional reconnaissance about the Montgomery flights, Root finally gained a proper perspective on the matter. On April 17, he apprised the Wrights privately:

> By the way, the California exploit of dropping down two or three thousand feet with a gliding machine, that is, if we have got the truth of the matter, seems to amount to something. The man who handled that gliding machine turning to the right and left, and then sliding down hill at terrific speed and going up again with the momentum required, must be somewhat of an expert; and if he can handle a gliding machine he probably is pretty well prepared to learn to run the same machine with a motor; and it looks to me as though you might have a rival somewhere on the Pacific coast.[22]

Meanwhile, Montgomery had refocused his attention on Baldwin. On April 7, 1905, he filed two lawsuits in the Superior Court of Santa Clara County, with John Leonard representing him in both cases. The first suit sought to restrict Baldwin from further exhibitions of *The California Arrow* and to enforce an accounting of the money he had earned through exhibitions of the airship in accordance with the contract that the two men had signed in April 1904; the second one asked for damages equal to $100,000 (the prize claimed by Baldwin at the aeronautic competition at St. Louis).[23]

Over time, Chanute became more appreciative of Montgomery's accomplishments. In mid-April he wrote to John to "most heartedly congratulate your success in the boldest feat ever attempted in gliding flight. You have my earnest wishes for continued success in the further experiments which you are about to undertake."[24] He asked for information on the basic design (overall weight and wing area) and performance (descent rate, average flying speed, and gliding angle) of *The Santa Clara,* adding, "This information will be required when you come to apply a motor."[25] He also expressed interest in John's intentions regarding patent protection. Although those who were close to Chanute knew that he was directly connected to the Wrights, that relationship was not common knowledge, and it was virtually unknown to anyone pursuing aviation in the western United States, including John Montgomery.

In a letter of April 20, John sent Chanute a photograph of the unnamed smaller 32-pound tandem-wing glider that he had built and flown at Cox Ranch (and which was now hanging in his shop), including details of its design and control. He closed by saying, "Please keep in confidence some of my arrangements." Upon receipt of this letter on April 28, Chanute responded, and in acknowledgment of the patent situation stated, "I thank you for your letter of 20th, and the photograph, which latter I will keep to myself for the present, as I infer that you are applying for a patent."[26] However, Chanute's friendship with and advocacy of the Wrights made him unable to resist revealing at least some detail about *The Santa Clara.* On the very same day that John sent his letter and photograph to Chanute, Chanute sent a letter of his own to the Wright Brothers, providing them with the basic dimensions and flight characteristics of Montgomery's design and informing them that John had been asked to present exhibitions for paying audiences, and that he intended to apply for patent protection.[27] For the next several months, Chanute continued his correspondence with John, gathering all the information that Montgomery was willing to divulge about the glider's design, construction, and performance. He routinely passed John's responses and letters to the Wrights for their review. While Chanute shared John's information openly with the Wrights, he never mentioned the Wrights in his letters to John at all.

John filed a patent application for his aeroplane on April 22, 1905, one week prior to his first planned public exhibition of the design. In early April, he also began construction on a duplicate of *The Santa Clara,* christened *The California.* April 29 was the annual "Feast Day" celebration at Santa Clara College, or as the Fathers called it, "St. Robert's Day." The celebrations that year were being held in honor of college president Robert Kenna's birthday and the fiftieth anniversary of the college. Following the glider flights in March, the college fathers and John's colleagues had encouraged him to use this planned public celebration as the opportunity for a public unveiling of *The Santa Clara.* The relationship between the Montgomerys and Santa Clara College was personal, and this day was a celebration not only of the school's anniversary but of the advancements in science that this close collaboration had helped to inspire. The exhibition was the college's opportunity to announce to the world that a respected member of the faculty, through years of careful scientific research and development, had achieved a significant breakthrough in aerial navigation. It was important to John that his mother be present at the public unveiling of his aeroplane. He wrote to Ellen in Los Angeles, inviting her to attend. "It will be the event of your life," he added, "and an opportunity afforded very few mothers."[28] On the bright, clear morning of April 29, a large crowd gathered on the college grounds. Representatives of the press, prominent members of the

Fig. 23. John Montgomery with *The Santa Clara* at Santa Clara College, April 29, 1905. This was the first large-scale public exhibition of controlled heavier-than-air flight in America. (Courtesy Santa Clara University Archives and Special Collections)

clergy, public officials, and members of the Montgomery family had come to witness the first public demonstration of *The Santa Clara.*

Frank Hamilton managed the preparation of the balloon as teams of students and bystanders held it in place with ropes. Daniel Maloney awaited the ascension, dressed in his traditional show outfit of tights and red silk shorts. John's cousin, Archbishop George T. Montgomery of San Francisco, showed his support by attending the event. *The Santa Clara* even received a benediction from Father Kenna prior to liftoff. John noted later in private correspondence, "Everything was ideal, filled with solemnity as Fr. Kenna in [the] presence of a throng with uncovered heads blessed the machine."[29] During the blessing, Father Kenna reiterated Archbishop Montgomery's plea that the aeroplane never be used for the purpose of war. The events of that day were best described by John himself, in a letter to Chanute sent the following day:

> Promptly at the time the balloon started, but owing to too much hot air and too small a rope, the balloon broke the main rope, started with a rush, thus depriving us the opportunity of getting a full set of photographs, and narrowly escaping smashing the machine to pieces. But it

withstood this very severe test, and reached a height of about 4,000 ft. and when cut loose behaved beautifully; every one of my instructions being carried out. The vast crowd was spellbound and swayed between feelings of intense fear and delight as they beheld the downward darting, swift whirling, with the machine tilted sideways more than 45°, only to pass gracefully again to its gliding movements. I had to reassure the people the aeronaut was following my instructions by pointing out and explaining the various moves. At one time he experienced the severest ordeal, as he told us when he reached the ground. The day was one of constantly changing light winds and when about at 1,000 feet high suddenly the wind seemed to be in all directions at once; and he thought he had struck a funnel of rising air: was carried up and seemed to drop, then rest almost motionless. Thence he glided around within easy reach of the appointed place, and one who saw the landing said he moved in a large circle moving very rapidly, then darted straight ahead, and with a sudden turn of the tail came to a stand and gently settled. He landed in a nearby open field I had pointed out to him the day before; as the town is filled with all kinds of electric wires, and it is dangerous to attempt any landing in the streets or within the college grounds. I intended to telegraph you but remembering I had made arrangements with the Press representatives, it was useless to do; it as you would receive full accounts in the morning papers.[30]

In a private letter to his brother James, who was still in the Yukon, John noted that "The crowd was spellbound and wavered between awful suspense and exultant cheering as the various moves were made. The day and its events are never to be forgotten."[31]

This flight was not only the first large-scale public demonstration of a heavier-than-air flying machine in the United States, but also the first (albeit unintended) recognition of a thermal by an American pilot, or perhaps any pilot in the world. That Montgomery correctly attributed Maloney's sudden rise in altitude to a "funnel of rising air" indicates his very thorough knowledge of the use of thermal soaring by birds, his knowledge of air currents, and the pilot's knowledge of what was happening in the air surrounding the glider while it was in flight.[32] Father Bell offered insight on the controllability of the craft: "When Maloney, by directing the movements of the wings and sails, put the machine at an angle of about 90 degrees to the earth and then in a graceful curve came circling about a second afterward, in the opposite direction, it was demonstrated that the machine is built on thoroughly safe principles."[33] At the critical moment when the balloon was cut free, Maloney recalled:

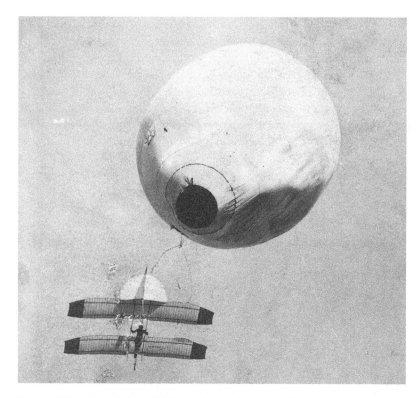

Fig. 24. A view from below of the Hamilton balloon and *The Santa Clara* during their ascension at an altitude of about 150 to 200 feet on May 21, 1905. (Courtesy Santa Clara University Archives and Special Collections)

Instantly the aeroplane settled on the air and there was not a drop of even ten feet. That sudden thrill of almost horror, something entirely new to me, came just at that moment as I had realized that the balloon had gone and that literally I had to fly for my life. . . . There was an indescribable thrill when I found myself launched on the air and about 4,000 feet above the earth. With nothing but those slender and flimsy wings to support me. I was actually flying and I believe that I am the first man on earth who has actually and successfully had that experience. . . . At first there was a little bit of a shudder, as I looked at the ground so far below me and I realized that I was sailing on a machine that weighted only 42 pounds and was so little that I could fold it up and carry it on my shoulder when walking on the ground, but the aeroplane so perfectly responded to every tug of the cords.[34]

Dennis J. Kavanagh, an associate professor at Santa Clara College, wrote: "All that is waiting for complete success is a power, in the machine itself, to

rise from the earth. This point has not yet been attempted by Mr. Montgomery, for the simple reason that to his scientific mind, it would seem
too much like greed, too much like ambitioning all or none and being left
empty-handed in the end."[35] Kavanagh's article included a quote that he
attributed to Alexander Graham Bell regarding the April 29 flight: "'The
Montgomery Machine will centralize all future experiments; with it subsequent investigations must begin."[36]

The public exhibition was reported in newspapers and journals across
the United States and in Europe.[37] Albert Zahm later placed it in perspective: "This daring performance amazed the world, and most of all, the
specialists who all along knew such a feat to be practicable."[38] Father Bell
acknowledged the significance of the event: "It must be understood that
Professor Montgomery in offering to the world the aeroplane, invented and
planned by himself, has solved the first part of a very intricate and complicated problem and has practically demonstrated what no one has done
heretofore—that an aeroplane built on thoroughly scientific principles can
be handled safely in the air at any height."[39] John Montgomery had succeeded in translating the seemingly impossible into practical accomplishment. At the conclusion of the flight, he was asked by a correspondent,
"After the success of your aeroplane, what gave you the greatest satisfaction?" He replied, "The things that gave me the greatest satisfaction were
mainly these; the solemn blessing of the aeroplane, and the presence of my
aged mother and of my brothers and sisters."[40]

Although John and his aeronauts made many successful flights from
hillside launches and several successful high-altitude balloon launches with
his flying machine, the flight of April 29, 1905, is the event that is most commonly cited in modern aviation history texts as Montgomery's contribution
to aeronautics. With the passage of time, this first public flight eclipsed his
prior accomplishments in aeronautics, electricity, and physics. The widely
publicized flights of March and April 1905 forever changed public perception regarding the reality of manned, controlled flight in heavier-than-air
flying machines. Man not only had flown, but he had flown long and high
over the skies of California.

CHAPTER 12

Tragedy

I regret that he has gotten into a wrangle with Baldwin and I fancy that in this and in the development of his aeroplane he is like the young bear who had all his troubles still to come.

Octave Chanute to Wilbur Wright regarding John Montgomery, April 16, 1905

In the first week of May 1905, Chanute visited Samuel Langley, Alexander Graham Bell, and Emile Berliner in Washington, D.C., and informed them that the newspaper accounts of Montgomery's flights were indeed genuine. At the same time, Amos Root, Chanute, and the Wrights were continuing to collect, share, and examine additional newspaper accounts of the April 29 flight. European colleagues also were sending newspaper accounts to Chanute, some of which he forwarded to John. In correspondence with his European friends, Chanute clearly acknowledged the important achievements of both the Wrights and Montgomery.[1] In addition, he viewed Montgomery's success as confirmation that the basic tandem-wing design, which Langley had also used on *The Aerodrome*, was sound. He noted this in a letter to Langley after Montgomery's flights:

> I am confident that if the machine had been launched, (and this was almost certain to occur upon the third trial) it would have made a good flight. This is evidenced by the experiments of Mr. Montgomery in California last March and April with a gliding machine shown by the photographs to have the same disposition of surfaces as your own, proportioned to about the same weight or loading per square foot, which aeroplane has repeatedly glided down, with gravity as a motive power, from heights of 2000 and 3000 feet, and alighted in safety.[2]

Newspaper accounts of the Montgomery flights piqued the interest of two Bay Area brothers, Malcolm and Allan Loughead, whose mother was the well-known Californian author and newspaper correspondent Flora

102

Haines Loughead. Allan and Malcolm were motorcycle and auto-racing enthusiasts and had been working in the auto repair business in San Francisco. During the late spring of 1905, they wrote to their stepbrother Victor Rudolph Lougheed (1877–1943) about the Montgomery flights.[3] Victor was a New York–based automobile consulting engineer who had authored technical publications and was the editor of *Motor Magazine;* in 1904 he became a founding member of the Society of Automobile Engineers (SAE). Like his stepbrothers, he soon caught the aviation bug and began to research and write about aviation. He initiated correspondence with John, and later became a strong advocate of Montgomery's achievements in aeronautics.

Meanwhile, John was receiving letters and telegrams inquiring about plans for future public demonstrations. Richard encouraged his brother to pursue this possibility as a source of additional revenue for research and development in aeronautics. As a result, John scheduled a series of public demonstrations while he and Maloney began construction on a Montgomery hot-air balloon, and he made improvements on the balloon release system that included a windlass. With Richard as their unofficial business manager, Maloney, who had been working on salary, was now offered a share of the profits.

Capitalizing on his association with the highly publicized Montgomery flights, Frank Hamilton went back to making balloon ascensions and parachute jumps at Idora Park. In what would become something of a tradition among Bay Area aeronauts, his wife, Carrie Clifford Hamilton, also made balloon ascensions and performed trapeze work and parachute jumps from balloons simply to encourage interest and keep up with rival acts. However, the recognition that "Professor Hamilton" received for his bold stunts paled in comparison to the fame accorded Maloney and the Montgomery exhibitions, and Carrie's demonstrations did little to change that.

On May 21, an exhibition of *The Santa Clara* was given at San Jose's Agricultural Park. As Montgomery and Maloney had not yet completed their new hot-air balloon, Hamilton was once again hired to supply and manage his balloon for this event. The glider *The California* was brought along as a backup to *The Santa Clara*. An admission fee of 25 cents was charged to help offset expenses (such as Hamilton's fee) and to reimburse the Santa Clara Valley Agricultural Society, which owned the park. Any surplus was left for John and Maloney to share.

Throughout the morning, the park grounds filled with 2,000 spectators as the fire pit was stoked to fill the balloon with hot air. As the crowd grew restless with anticipation, Hamilton gazed out at the swelling masses,

Fig. 25. Maloney ascending with *The Santa Clara* at San Jose's Agricultural Park, May 21, 1905. *The California* was also brought along as a backup. (Courtesy Santa Clara University Archives and Special Collections)

Fig. 26. Later in the morning of May 21, Maloney ascends in *The California*. Note *The Santa Clara* set aside in the trees. (Courtesy Santa Clara University Archives and Special Collections)

resentful of the money being made by others. He entered the ticket office and announced to Richard that his balloon would not be used for the ascension unless he received an advance payment of $500.[4] A lengthy argument ensued between Hamilton, John, Richard, and Carrie, Hamilton's wife, during which Richard eventually ventured outside the ticket booth, only to discover that someone had opened a rear gate to the park and was collecting his own admission fees! Hamilton's extortion demand represented essentially their entire net income. After being apprised of the situation, Maloney insisted on making the flight regardless of compensation.[5] Angered by Hamilton's audacity but realizing their obligation to appease the large and increasingly restless crowd, Richard capitulated and handed the gate receipts to Hamilton, who promptly deposited the sum into his wife's purse. In doing so, he doomed his relationship with Montgomery.

As final preparations were made to tether *The Santa Clara* to the balloon, John asked Hamilton whether the ropes (which had seen heavy use in recent weeks) were up to the task. Hamilton replied curtly: "I know my business!" Richard later noted, "I presume he realized he would be discharged after this flight was over."[6] Yet as *The Santa Clara* began its ascension, the rope connecting the glider to the balloon broke at an altitude of only 150 to 200 feet. Maloney reacted immediately and yelled down to the crowd to get out of the way as he skillfully maneuvered the glider, avoiding trees and the grandstand as well as groups of spectators to make a safe landing on the racecourse about 100 yards from the point of the balloon ascension.[7]

The spectators jeered and hooted, some demanding a return on their entrance fee. In all the excitement, *The California* was left unguarded at the periphery of the exhibition area. In order to avoid disappointing the crowd, the balloon was filled once again, and the backup glider was brought out and connected to a newly inflated balloon. As *The California* ascended, it became entangled with two guide ropes. John Montgomery yelled to Maloney to stay with the balloon as it cooled rather than risk releasing the glider. As a result, Maloney drifted three miles to the south before landing in the glider while it was still attached to the balloon.

During the slow descent, Maloney discovered that some of the bolts for the pulleys and other key elements of the control system had been loosened, or even removed entirely. After a safe landing, he relayed this information to John, who concluded that the glider had clearly been tampered with in a deliberate attempt to endanger Maloney's life. As all this was unfolding, some in the crowd began saying that Maloney was a coward or that the aeroplane was a "fake." John and Richard drove the wagon to Maloney's location, packed up the equipment, and made the trip back to Santa Clara College.

John found the attempted sabotage of *The California* disturbing and pondered the identity of the culprit. Aside from the Montgomery brothers, Maloney, Hamilton, and perhaps Father Bell, no one but Baldwin had such familiarity with the control system. Shortly after returning to Santa Clara, John was informed that Baldwin had indeed been spotted at the exhibition of May 21. The park was equidistant from Santa Clara College and Randall's Cyclery, a location that Baldwin frequented. It was also reported in the newspaper that Baldwin had arrived in San Jose from Los Angeles a week prior to Montgomery's exhibition, and that he planned to fly *The California Arrow* at the very same park.[8] John therefore concluded that Baldwin or an associate of Baldwin's had been the responsible party. (In retrospect, it is possible that Hamilton, or a Hamilton associate, might have tampered with *The California,* although the identity of the true culprit remains a mystery.) Montgomery's local exhibitions raised the bar in terms of aerial feats. As a consequence, northern California–based aeronauts and balloonists, saw a reduction in public interest for their own demonstrations. Jane Montgomery alluded to the impact of this situation of John's status with his aerial peers: "the parachute and balloon men were all against him because he was injuring their work."[9] With events unfolding rapidly in the spring of 1905, the impact of John's work on his peers was lost on John himself. Within days of the exhibition, he traveled to San Francisco on a lead that Baldwin had been spotted, but Baldwin eluded him. Over the intervening days, John took his case directly to the public for their scrutiny through a series of newspaper interviews in which he chronicled the events that had led to the dispute.[10]

After weeks of ignoring John's statements in the press about their falling-out, Baldwin was swift to respond to the very public accusation that he had tampered with *The California*. In a newspaper interview on June 1, he denounced the charges as "a pack of lies!" "The aeroplane invented by Professor Montgomery," Baldwin lamented, "is really nothing better than a disguised infernal machine, certain to kill any man who shall try to operate it at an appreciable altitude." He said that he had abandoned the contract because the Montgomery aeroplane had failed to demonstrate that it could do what John claimed it could do.[11] Ignoring Montgomery's successful exhibitions, Baldwin further noted, "Instead of being satisfied in that respect, I was led to realize before the first test of the machine was made, that to attempt to navigate the aeroplane at any elevation much higher than the treetops would be foolhardy."[12] In response to a question about the propeller design for *The California Arrow*, he commented, "There is absolutely nothing new about *The Arrow* that came directly or indirectly from Professor Montgomery, which I shall be glad to demonstrate to anybody interested.... It was only toward the end of our relations that I began to suspect that he was envious of what he supposed were my financial returns from the *Arrow* exhibitions.... He is a scientist and, ought to know where he gets off. I am not a scientist and don't pretend to be. I am merely a showman."[13] Baldwin made it clear to the press that he intended to sue Montgomery for libel.

Baldwin's interview flew in the face of his own previous public statements, in which he had credited his propeller design initially to himself, then to both Montgomery and Father Bell, then to only Father Bell, and subsequently, after Bell's public denial, once again only to himself. Despite the many contradictions, as late as November 1905, Baldwin again wrote that he had taken "a special course in physics at Jesuit College, Santa Clara College, Cal, to aid [Baldwin] in the problem of aerial navigation."[14]

Baldwin's comments about Montgomery caught the attention of August

Greth, whose public retort was published on June 2.[15] Greth supported Montgomery and noted, among other things, that over the course of his sporadic association with Baldwin, he had been alienated by Baldwin's frequent "double-dealing," which had led to their eventual parting in late April 1904. The *San Francisco Examiner* article noted that Greth had "catalogued Baldwin's statement that Professor Montgomery's aeroplane was impracticable and dangerous as an opinion of no value whatever, the latter, being an undoubted scientist, the former, being a charlatan." Greth added the tongue-in-cheek comment, "the best that can be said of Baldwin is that he would shine as a circus clown or tightrope dancer."[16]

Also on June 2, in San Francisco, Baldwin convinced Police Judge George H. Cabaniss that he had been traveling by train from Portland on May 21, the day of Montgomery's exhibition at the Agricultural Park in San Jose. Because John had implicated Baldwin as the saboteur without direct evidence, a warrant was issued for his arrest. He received a phone call from the sheriff's office that evening notifying him of the order. The following day, he made an appearance with his attorney at a local sheriff's office at San Jose, paid $50 in bail, and returned to Santa Clara College. He defended his position in an interview with the *San Jose Mercury News* that afternoon.[17]

After successful trials at Santa Clara with a new hot-air (or "smoke") balloon, John decided to hold a free public exhibition of glider flights on June 25 at the Emeryville Racetrack just north of Oakland. On the day of the exhibition, gusty winds arose, forcing everyone to wait several hours before the balloon could be filled. As part of the inflation procedure, a helper named Nels Larsen positioned himself inside the balloon to hold its base open while the constant burning of kerosene-soaked wood filled the interior with hot, smoky air. The expanse of cotton cloth that formed the balloon itself grew larger as it surged in the gale. Dozens of men were dragged to and fro in spite of their combined efforts to steady the growing balloon. Then an especially violent gust of wind created a large tear in the fabric, and in a matter of seconds it was reduced to "a dangling mass of sooty rags," with nothing but a huge cloud of inky smoke remaining.[18] Larsen had been overcome by poisonous gases inside the balloon and was found unconscious. John stayed with him until a doctor arrived, after which he came to and was fine.

Another demonstration of *The Santa Clara* was arranged for San Jose's 4th of July celebration at the Agricultural Park. Veteran balloonist Park Van

Tassel was hired to perform a balloon ascension and parachute jumps for the event, and also to use his hydrogen balloon to lift *The Santa Clara* to altitude. He had been giving solo exhibitions following the death of his wife Jeanette in a parachute jumping accident in Bangladesh in 1892. After Van Tassel's successful jump, he supervised the refilling of his balloon. When the newly expanded balloon was released, however, it failed to lift both Maloney and the aeroplane (a combined weight of roughly 188 pounds). An exasperated Maloney grasped the base of the balloon, concluding that Van Tassel had shortchanged them on the amount of hydrogen gas (an expensive commodity at the time). The crowd booed, proclaiming Maloney to be a "faker" and a coward. Out of frustration, he pleaded with John to allow him to launch from the top of a nearby building. Although the glider had previously been launched from hilltops by using a derrick, and foot-launched from a slope, self-launching from a building was an entirely different matter. John refused to allow it. That a veteran balloonist like Van Tassel was unable to fill the hydrogen balloon enough to lift 188 pounds was as perplexing as it was suspicious. This incident, coupled with Hamilton's extortion at the May 21 event, convinced John that the California-based balloonists viewed exhibitions of the Montgomery aeroplane as a threat to their livelihood, and that they were hoping to defeat his efforts. From that point on, he no longer relied on balloonists.

On July 15, another aerial exhibition was given at the Agricultural Park in San Jose. It included the first joint exhibition of a heavier-than-air craft (*The Santa Clara*) and a lighter-than-air craft (George Heaton's dirigible *The California Messenger*) in the western United States. In the morning, *The Santa Clara* was lifted by balloon to an estimated elevation of 3,000 to 4,000 feet. However, when the balloon ceased its ascent, Maloney was unable to cut away from it. A rotten guide rope had brought a huge knot in contact with a block and pulley, thwarting every effort on his part to cut loose.[19] Yet again he was forced to descend with the glider still attached to the balloon as it cooled. As the rate of descent increased, Maloney used the glider's controls to help guide it down. A second ascension was not possible because of a horse race scheduled to be held at the park later that day. Heaton's dirigible made two successful flights for the crowd, but Heaton later realized, much to his disappointment, that a majority of the spectators had chosen to view the exhibition from outside the park rather than pay the entry fee of 25 cents. However, the crowd was largely satisfied with the event because of Heaton's display. Despite being frustrated by the series of failed launches, John felt that the setbacks could be overcome with experience, refinement, and additional airtime. Maloney, however, took the criticism more personally, as a direct challenge to his skill as an aeronaut.

Maloney had maintained a strong friendship with some former fellow members of the League of the Cross Cadets, a Catholic temperance organization for young men. The league's annual encampment was scheduled to take place at Santa Clara College on July 18, 1905. Maloney saw this as his opportunity to restore his reputation with his peers on his "home field," and a demonstration flight was arranged specifically for their benefit. The morning of July 18 presented calm, warm, ideal conditions for flight. Excitement among the Cross Cadets mounted as the balloon was inflated and the glider was positioned and attached to the ropes. The tenders eased the balloon upward slowly by letting the holding-down ropes slip gently through their hands. This critical step in the launch procedure had been the source of many previous problems because of the distractions caused by the launch commotion. Maloney signaled to the tenders to let the balloon up gently until the cable from the balloon to the plane was taut. When this was done, he shouted "Let go!"

Just as the glider lifted from the ground, several of the young men assisting with the launch noticed that it had been damaged by a heavy rope as the balloon made its quick liftoff. The damage was behind and out of Maloney's view.[20] John described the subsequent events in a letter to Chanute:

> As the balloon was ascending with the machine, one of the ropes that was being drawn from the balloon fell with a snap and wrapped around the wing . . . bending it down violently. As the front arm of this wing is supported by wires leading to the tower [a cabane] . . . the latter was broken. This tower not only supports the front arms of the rear wings but also supports the pulley holding the rope controlling the tail.[21]

Cross Cadet Robert L. Defolco called out to Maloney that the machine had been damaged. It remains unclear whether Maloney actually heard any of the voices from below in the confusion of launch. As the balloon ascended, some of the Cross Cadets cheered, and Maloney called out "Hooray for Professor Montgomery!" But Montgomery, Bell, Defolco, and the others who had been holding the balloon watched anxiously, wondering whether Maloney was aware that the rear cabane and wing were damaged. John later noted, "When the balloon was at its highest point the machine remained with it for some time and several present judged this to be a sign that Maloney considered it would not be safe to cut loose."[22]

Fig. 27. Glider exhibition at Santa Clara College, July 18, 1905. The moment of liftoff required a complex interplay between those tending to the balloon and another set of tenders for the glider in addition to the windlass system used by John for incremental control of the balloon at the moment of liftoff. Immediately after this photo was taken, a rope dangling from the balloon damaged the rear wing and the rear mast that supported *The Santa Clara*'s control system. (Courtesy Santa Clara University Archives and Special Collections)

It appears that upon inspecting the control system, Maloney realized that the rear cabane was compromised, potentially affecting the operation of the tail. He considered his options: he could stay attached to the balloon and descend with it as it cooled, or he could attempt flight despite the damage. The crowd of friends below waited in excited anticipation for his glide. In a fateful decision, Maloney released the glider from the balloon. John later noted:

> It parted from the balloon and glided beautifully about a thousand feet. But it did this because the tail remained as I had adjusted it. . . . But when Maloney lowered it to gain speed he failed to draw it back, and as the machine gained in speed the rear wings commenced to flutter violently and the machine, heading down, turned three forward somersaults and settled on its back.[23]

The excessive speed and forward somersaults had generated loads that far exceeded the capacity of the structural design. Charles Leahy, one of the students who had held the ropes prior to ascension, later recalled the

moment when the crowd realized that they were witnessing a tragedy un-
fold: "When we saw the rear wings fold up, the glider went into a series of
somersaults as it was coming down. Montgomery said, 'Oh my God!' Father
Bell took off his biretta and gave the sign of the absolution."[24] A low groan
went up among the spectators; birettas and hats were lifted, and many a
prayer was said.[25] The dive was at a slight angle, and many of the witnesses
were hoping that Maloney might be able to pull out of the fall. However, at
about 150 feet above the ground, the wings collapsed completely. The rapid
final plunge sent Maloney and *The Santa Clara* into the yard of Eberhard's
Tannery. The workers there, who had stopped to watch the flight, surmised
that to the very end, Maloney was trying to direct the crippled craft to a
large pile of tanbark to soften the impact. The aeroplane instead fell onto
the main entrance road of the tannery.[26]

A group of workers and Cross Cadets removed Maloney from the
wreckage and carried him to the college infirmary, where local doctors and
a nurse attended to him. His breathing was shallow and his pulse fluttered,
yet there was no sign of blood. Arriving at his bedside, John moaned in
anguish, "Oh Mother of God, why did this happen? I prayed so hard for
safety."[27] Maloney succumbed to his injuries within thirty minutes.[28] John
was devastated. "The scene around his deathbed was a sad one," a local
newspaper noted. "Maloney had endeared himself to every one connected
with the college by his modest, unassuming manners, his bravery and his
indomitable will. Prof. Montgomery was heart-broken. He walked the halls,
moaning and wringing his hands in anguish over the terrible disaster which
had caused the death of his true friend and fellow-worker."[29]

At the coroner's inquest, several Cross Cadets, including Robert Defolco, testified that Maloney had repeatedly expressed a desire to reclaim
his good reputation. During a visit with some of the Cadets on the prior
night, he had said that he would put an end to the accusations that he was a
"fake": he would "fly or break his neck."[30] Maloney was no doubt aware that
if he had shared those thoughts with John, he would not have been allowed
to make further flights. However, having put his faith in the machine, his
aeronaut, and God, John did not anticipate Maloney's reaction to the public
scorn. The coroner's jury absolved Montgomery of any responsibility for
the accident. It was the consensus of the majority who provided testimony
that Maloney's zeal to right his reputation had contributed to the clouded
judgment that resulted in his reckless decision to release from the balloon.

Maloney's death was the world's third fatality involving a successful
aeronaut of heavier-than-air flying machines (following the earlier deaths

of Otto Lilienthal in Germany and Percy Pilcher in Great Britain) and fueled the public's perception of flying machines as dangerous "man killers."
Coming on the heels of Baldwin's assertions regarding the "infernal" design
of the Montgomery tandem-wing gliders, Maloney's unfortunate fate could
not have been more ill-timed. The accident made national headlines. The
front page of the next day's *New York Times* featured an article headlined
"Aeroplane Fell 3,000 Feet," attributing the cause of the accident to a structural problem rather than Maloney's piloting or the glider's design.[31] Newspapers across the country printed similar accounts of the accident.

John never considered the injury or loss of an aeronaut to be an inevitable consequence of his work. His exhaustive experimentation with
unmanned models and his desire to study aeronautics had reinforced his
belief that the operator's safety was of primary concern. This philosophy allowed a pilot to successfully control the aircraft after only a modest amount
of training. The problems encountered with Montgomery's gliding demonstrations had everything to do with the tenuous nature of the balloon-
assisted launch method.

Daniel Maloney's resolve, his faith in the Montgomery flying machine,
and his willingness to patiently follow a script of aerial maneuvers had
provided him with the opportunity to "cleave the air at will." His primary
contribution to aviation was human-controlled gliding flight, and on a
scale that had never before been achieved, thereby increasing the public's
awareness of the promise of aerial navigation. Even today there is a general
ignorance of the purpose and scope of his flights. Although he is usually
given a footnote in aviation history for his flights in the spring of 1905, he
deserves a great deal more credit for being the world's first high-altitude
pilot. Montgomery's contemporary Albert Zahm noted that Maloney's records for high-altitude flight were not exceeded until 1911, and that was by
a powered aircraft, only a few years after powered flight had become rather
commonplace.[32]

Upon hearing of the accident, both Chanute and Root notified Wilbur
Wright immediately, having drawn similar but independent conclusions
that the accident was the result of structural damage incurred during liftoff
rather than any design flaw. Wright, however, refused to accept the obvious.
Instead, he asserted in a letter to Chanute, "In the accounts I have seen of
the Montgomery catastrophe, there are so many more facts going to prove
that the trouble was not due to a broken spar than there are in support of
that theory, that I cannot believe the trouble due to that cause."[33] In an effort
to be objective on the matter, Chanute made direct inquiries to some of the

Jesuit priests who had witnessed the accident and was later satisfied that his own impressions were correct.[34] As the weeks passed, an alternate explanation for the accident failed to materialize in the press, so in early August, the Wrights provided their own tangential explanations in a letter to Chanute. Entrenched in denial and ignoring all details regarding prior flights and the structural damage that was clearly noted in the accounts, Wilbur offered increasingly elaborate theories, with negative connotations that were largely meant to discredit Montgomery's design, his theories, and any of his prior success. Several months after the accident, he ended one letter with a surprising statement: "the tragic death of poor Maloney seemed the more terrible to me because I knew it was coming and had tried in vain to think of some way to save him. I knew a direct warning would tend to precipitate rather than prevent a catastrophe."[35] Despite Wilbur's effort to sway others' opinion, Chanute's first published account, which appeared some months later, attributed the cause of the accident to structural failure.[36]

On the same day that Maloney met his fate, Baldwin's attorney appeared before Judge Cabaniss at San Francisco and requested that the charges of libel against John Montgomery be dropped. He and Baldwin had come to the realization that John had been misrepresented in the press. The *Los Angeles Herald* reported, "It was admitted by both sides that the professor had been misquoted."[37] The case was accordingly dismissed. It is unknown whether the decision to withdraw the lawsuit was in any way related to the events of the morning of July 18, or whether it was purely coincidence. Although John was relieved that the suit was over, the news was small compensation for the tragic loss of his friend.

CHAPTER 13

Wings in Tandem

F ollowing the Maloney tragedy, John sought solace. He stayed at the family home in Oakland, and then with Peter Cox's family at the Pina Carta Ranch in the hills near San Juan Bautista.[1] On July 20, 1905, he received a letter from an Arizonan named C. E. Warren who was well connected with the mining industry in the silver belt of central Arizona. Warren had written to encourage John not to become disheartened, and to offer his assistance.[2] He suggested that the Montgomery aeroplane might be useful for aerial reconnaissance in mining regions.

After recuperating for three weeks, John moved forward with his work. He purchased material for the construction of a new glider and hired Robert Defolco as his new aeronaut. Little is known of Defolco other than that he had met Maloney when they both had appeared in trapeze and parachute exhibitions prior to 1905 while both were also involved in the League of the Cross Cadets. After Maloney's death, Defolco expressed an interest in serving as his friend's replacement. John had new plans that called for structural improvements to strengthen his glider design. The remains of *The Santa Clara* were stored, and *The California* was suspended from the rafters in his laboratory. John and Defolco constructed the new tandem-wing glider with cabanes composed of wood and aluminum.[3] Aluminum was used for the beams that braced the leading edges of the wings. After experimenting with different types of silk for wing covering, John selected a Japanese variety, applying oil to reduce its permeability. A friend, Philomena Ruth, sewed pockets for the wing ribs so that they were "glove tight," allowing for only slight movement of the ribs when they were being warped by the control system.[4] A short spruce beam was substituted for the control stirrups used previously to independently depress the rear portions of the wings for control.[5]

Montgomery was invited to exhibit his aeroplane at the International Exposition planned for April 1906 in Milan, Italy.[6] The intended purpose of the exposition was to showcase the latest developments in both communication and transportation. However, other events would intervene. John also received offers from two syndicates in the eastern U.S. for the purchase of Montgomery aeroplanes, but he considered the requests to be premature, so he turned them down.[7] It was reported in the press that he planned to exhibit his aeroplane as part of a national tour, to include Reno, Kansas City, Salt Lake City, Cheyenne, Chicago, Cincinnati, Baltimore, and New York City.[8] This was an idea that he never followed through on.

In late August, John received an invitation from the executive board of the Sacramento State Fair to exhibit the Montgomery aeroplane as a key example of California's contribution to successful human-controlled flight. The fair would also mark Sacramento's fiftieth anniversary as California's capital. With encouragement from family, friends, and colleagues, John agreed. He arranged for a flight by the newest glider, *The Defolco*, combined with a display of *The California*, noting in an open letter to the press: "I am desirous of giving an exhibition at this time and in this place for the reason that it is the fair of my native state; my invention is a California product, and I wish the sons and daughters of California to witness its workings."[9]

Upon arrival in Sacramento on September 3, *The California* was placed on static display for public viewing at the Agricultural Park (Sacramento). Private tests with *The Defolco* were made at the fairgrounds, with Robert Defolco serving as the aeronaut. The new windlass was evaluated thoroughly by raising and lowering a series of balloons. A public glider flight was planned as the prime event for Governor's Day on September 7.[10]

Training flights were made on September 6, during which Defolco was jostled in the craft by strong winds that led to partial damage. The repair meant that John missed the scheduled exhibition date on September 7. On September 8, *The Defolco* was ready for a public demonstration. With only a few days of training under his belt, Defolco made a balloon-assisted glider launch from an altitude of 500 feet, at the end of which he landed lightly on his feet near the point of ascension.[11] No grandiose maneuvers or extended glides were attempted. That Defolco was able to accomplish this public flight with such success after such a short training program is a testament to the stability of the Montgomery design. However, in light of the well-known reports about Maloney's long, high-altitude flights with intricate maneuvers, some in the crowd complained about Defolco's comparatively timid performance. The public reaction was that this admittedly successful flight was nothing extraordinary, but merely "more of the same."

Had the members of the State Fair committee reviewed the state of American aeronautics at the time, they might had realized that in the period 1903–1905, California-based pioneers (Montgomery, Greth, and Baldwin) had made significant contributions to the advancement of American aeronautics in both classes: lighter-than-air and heavier-than-air. For all of Baldwin's shortcomings, it was he who had acted as a central catalyst in successfully translating Greth's and Montgomery's ideas into successful dirigible technology, and simultaneously (and inadvertently) sparking Montgomery to go public and establish his place as a successful pioneer in the field, which had led to increased public acceptance of human-controlled flight. His exhibitions and the resulting articles about them presented the conceptions of a California brain trust. By this time, Baldwin had already shared his propeller theory and dirigible technology with some of America's first generation of aviators, including A. Roy Knabenshue, Lincoln Beachey, James C. "Bud" Mars, and soon, Glenn H. Curtiss. In so doing, he gave critical impetus to some remarkable careers in American aeronautics. Additionally, Jeanette Van Tassel, Carrie Clifford (Hamilton), and Georgia Heaton had paved the way for women in the art through their bold efforts in lighter-than-air flying machines and other aerial feats. These trailblazers changed public perception of the role of women in a traditionally male-dominated field.

However, such feats were not without substantial risk. In 1905, the *Oakland Tribune* noted that in the period since 1885, a total of 2,051 accidents had occurred during exhibitions of lighter-than-air flying machines.[12] One such incident, which involved Georgia Heaton, was particularly noteworthy. On September 11, 1905, shortly after Defolco's flight in Sacramento, Mrs. Heaton agreed to have her photo taken in *The California Messenger,* to help bring attention to her role in the invention. As the photographer prepared to take the shot, the airship managed to release itself from the several men tending the ropes and started upward on an uncontrolled ascent. Heaton kept her cool. She first tried to steer the ship downward, but without success. The dirigible took a meandering course hundreds of feet aloft. In an attempt to release some gas from the balloon, she tore a hole in the silk, and inadvertently in the supply tube as well. But the gas, concentrated in the upper half of the dirigible, would not escape.[13] She attempted to coax the dirigible into an angle that would allow gas to escape by climbing out onto a bamboo mast that protruded from the rear of the ship, "gripping like death" as she hung on the mast at the deflated end of the balloon. She then swung hand over hand back to the gondola. As she rushed back to the front of the gondola, the ship finally made its rapid descent, returning her safely to earth.

From May to October 1905, the Wright Brothers forged ahead with their latest powered flying machine, the *Wright Flyer III*. In late May they incorporated several design changes from the 1904 *Wright Flyer II*, including decoupling the wing-warping controls from the rudder so that all of the controls could be operated independently.[14] After further flights in June, July, and August, they reconfigured the tail assembly, and as Wright biographer Tom Crouch noted, "the aircraft that emerged differed significantly from a few weeks before."[15] Improved results were apparent rather quickly as experiments continued at Huffman Prairie near Dayton, Ohio. The flights of the *Wright Flyer III* in September and October 1905 represented a significant improvement over those accomplished in the fall of 1904. The occasional reports of the Wrights' success that had been leaked to the press in the previous fall had often been received with incredulity and disbelief.[16] Nevertheless, the brothers had finally achieved their goal—they could deliver a machine for sale, specifically tailored to meet practical requirements such as those of the U.S. government for use in military reconnaissance.

The patent process itself had the potential for large reward if proper strategy was used. For instance, in the famous case of George Selden's internal combustion engine, the inventor submitted change after change to the U.S. Patent Office regarding his original application, year after year. "By the time Selden's calculated delays had run their course," one author noted, "he had withdrawn his original nineteen bases for a patent and replaced them with some one hundred modifications and amendments based on a year-to-year progress of other inventors. Hence, he was able to absorb the latest technological advances into his original worthless patent."[17] When the Wrights were attempting to establish the broadest claims possible for their aeroplane patent, they also appreciated these nuances.

By mid-1905, there had been more than two years of review and communication with the Patent Office in the Wright Brothers' case, with the patent examiner offering some objections based on prior art.[18] On August 17, the Wrights' patent attorney wrote to the Patent Office requesting that previous submissions be canceled and replaced with a whole new set of file wrappers and specifications.[19] One critical change to their application pertained to the helical twisting of the wings. The Wrights attempted to broaden the description of their method to include any line of flexure at any point between the front and the rear of the airfoil (as would be the case with an aileron or flap).[20]

The concept of a point of flexure at an intermediate point on the airfoil, and in particular near the rear of the wing, was clearly similar to the Montgomery design principle from the 1880s, the details of which the Wrights had been exposed to through published and unpublished photographs and articles forwarded to them by Chanute and Root. This new specification appeared to be an attempt to expand the scope of the Wright patent application under special circumstances, beyond those of Montgomery.[21] The Montgomery patent application was in review at the patent office at the very same time, and by the very same patent examiner, William W. Townsend. The Wrights' new material was rejected after Townsend concluded that it was "new matter" and not admissible with respect to their prior specification.[22] The concept was subsequently withdrawn by the Wrights on their revised application of December 6, 1905.[23]

The Wrights' patent also included repeated reference to their airfoil as "a normally flat aeroplane." Their patent attorney, Harry Toulmin, had inserted the phrase during one of their early revisions to the specification, a fact that Wilbur later complained about in 1909 in private correspondence with Toulmin: "Why was it necessary to insert the underlined words. . . . it apparently absolutely ruins the claim."[24] Publicly, however, they continued to offer conflicting statements about the relevance of this particular wording. The timing and nature of the changes to their patent specification provide strong evidence that their intelligence gathering was used in an attempt to expand the scope of their patents, and specifically to maintain an advantage over their competitors.

Following the Maloney accident, exhibitions of heavier-than-air flying machines ceased in the San Francisco Bay area. Sensing an underutilized market niche, Frank Hamilton, who had continued to perform parachute drops and acrobatic stunts from tethered hot-air balloons at Oakland's Idora Park during the spring and summer of 1905, hired David Wilkie to help him construct a glider of their own. Wilkie was an Oakland native with prior experience in automobile and motorcycle racing. In 1905, he served occasionally as a reserve aeronaut to Georgia Heaton in *The California Messenger* and had been exposed to *The Santa Clara* at an exhibition given at San Jose.

On the strength of prior exposure to Montgomery's gliders, Hamilton and Wilkie constructed a replica of *The Santa Clara* and christened it *The Hamilton Aeroplane*. Hamilton then announced his intention to exhibit the new glider in Oakland's Idora Park.[25] Descriptions of the aircraft read exactly like a description of *The Santa Clara*: "It has four wings and a tail; the dimensions of the wings are twelve feet by four and the machine is made of muslin, hickory and spruce."[26] When Hamilton was questioned by an observant newspaper correspondent about the resemblance to Mont-gomery's design, he retorted that it had "many distinguishing features," yet these remained unspecified.[27] John took notice of the articles and re-sponded immediately by obtaining a temporary injunction, issued by Judge Waste on December 9, prohibiting any exhibition of *The Hamilton Aero-plane*.[28] On December 22 he went to court, and the injunction was made permanent.

During the fall of 1905, Octave Chanute prepared an article for the *New Standard Encyclopedia* in which he profiled the leading experimenters in heavier-than-air flight. He included himself in this list, along with Li-lienthal, Montgomery, Pilcher, the Wright Brothers, and others. The article was reserved for later inclusion in Hermann Moedebeck's *Pocket-Book of Aeronautics*, which was published in 1907 and became a definitive refer-ence book on aeronautics in the period. At this time John received a copy of James Means's 1897 *Aeronautical Annual* as a gift from Chanute. The *Aeronautical Annuals* series (published in 1895, 1896, and 1897) were es-sential reading for anyone interested in aeronautics after the 1893 Chicago conference, as they brought enthusiasts up to date on the latest data on the subject. Comments made by John at the time show that he had remained unaware of these publications until they were brought to his attention by Chanute. He was also surprised to learn in the *Annuals* that Chanute had conducted glider experiments at Miller Beach in 1896.[29] Despite his own unfamiliarity with the work of others in aeronautics, Montgomery's accom-plishments were recognized by others. In December 1905, he was invited to serve as a founding member of the Aero Club of America, an honor af-forded only to the leading experimenters in aeronautics.[30] The Aero Club was organized to promote aviation in general, and later became the Na-tional Aeronautic Association, an organization that is still active today.

After the injunction against Hamilton and Wilkie was issued in Janu-ary 1906, Wilkie approached John to voice his regret over his association with Hamilton. During a series of meetings between the two men, Wilkie expressed an interest in serving as one of Montgomery's aeronauts. John agreed and began instructing him on the theory and workings of the glider *The Defolco*. In the interim, Robert Defolco took on other obligations, but he returned for a brief period in 1906. Wilkie later confirmed that John "was confident his glider would take a motor but he wanted to obtain inher-ent stability."[31]

John sought a new location for private glider instruction at a hillside off Magdalena Avenue near the Jesuit-owned "Loyola Corners" property in Los

Fig. 28. David F. Wilkie with *The Defolco* preparing to make a launch from a suspended rope and derrick atop a hillside off Magdalena Avenue near the Jesuit-owned "Loyola Corner" property in Los Altos Hills, California, February 1906. (Courtesy National Air and Space Museum, Smithsonian Institution, SI 75-4250)

Altos Hills, California. There he identified a hillside that could be used for private tests and training. John and Wilkie were assisted by Defolco in constructing a derrick similar to the one used at Cox's ranch in 1905 for flight testing. Two telephone poles were installed at a distance of 40 feet apart. A cable was strung between the tops of the poles, and a pulley was used from the cable to hoist the glider to a height of roughly 20 feet for release in both unmanned (weighted) and manned configurations. Seventeen successful manned derrick-launched glides were made.[32] Glides were also made by foot launch down the hillside.

In January 1906, John wrote to Chanute, lamenting the overall difficulty and lack of progress he was experiencing because of characters such as Baldwin and Hamilton as well as the difficulty of finding qualified assistants with sufficient aptitude in engineering. Chanute offered his assistance but admitted that he knew nothing about the inner workings of Montgomery's designs and systems.[33] In a subsequent letter, Chanute revisited John's theories in aeronautics, and in the process revealed that he had not truly attained a full understanding of the theories required for a critical analysis:

> [I] am glad to learn that you are resuming your experiments. I hope that they will satisfy you as to the truth of the deductions which you have drawn from the theory of the rotation of the air currents, which I confess

I do not understand very clearly. That such rotations exist I well know and I have pointed out in some of my writings, but I do not feel sure, (not knowing your construction) that you have drawn correct inferences as to the best way of meeting them.[34]

Chanute had known about John's rotation theory of lift for a decade; it had been revealed publicly during John's lecture to the Southern California Academy of Sciences (1897), then published in the *Aeroplane* pamphlet (May 1905) and also in *Scientific American* (November 1905). It appears, however that with the increased public interest in Montgomery's work, Chanute was willing to be more objective in attempting to understand Montgomery's analysis.

After Wilkie completed training in late February 1906, the time had arrived for higher-altitude balloon launches in order to refine his skill through longer glides. On the morning of February 22, *The Defolco* was assembled at Idora Park in Oakland. During preparations for the ascension, word spread quickly, and spectators gathered. As the balloon was inflated, Wilkie checked the wires and control system, chatted with members of the audience, and then double-checked the aeroplane. He spoke briefly with reporters before the flight and offered a portent of things to come: "Last night I dreamt that after I got up with the aeroplane I became nervous and was afraid to cut loose but that I finally did so, and the machine fell. It seemed that I could not control it, and that I kept going faster and faster toward the ground. I am not superstitious though, and in spite of the dream I am going to make the ascension."[35]

At the time of the ascension, Wilkie was instructed to cut loose when John fired a starter's pistol as a signal (a planned altitude of 500 feet). Wilkie, however, ignored the signal because he mistakenly thought that he was situated too far aft on the longeron. While he repositioned himself, his clothes caught in the clutch controlling the tail and lowered it from its usual neutral position. All the while, the glider continued its ascent until it reached an altitude of 2,000 feet. With Wilkie's position in the glider too far forward and the tail depressed, cutting loose from the balloon caused the machine to begin a steep descent with a series of rapid spiral turns. Becoming frightened, Wilkie removed his feet from the stirrups that controlled the warping system, and, clinging to the riding bar, he let the machine proceed on its wild course. Out of control, the rear wings swayed like the arms of a balance, turning the machine rapidly from side to side. At about 300 feet, Wilkie managed to turn the elevator upward, slowing the machine's descent. However, his forward position in the machine brought the glider to an abrupt impact in a marshy area.[36] Although Wilkie was dazed, bruised, and suffered a cut through one nostril, he was able to stand immediately after landing and inspected the glider, noting that the damage was primarily to the front wing and cabane. He chatted with bystanders, assuring them that he was okay. His bruises and cuts did not keep him from his night shift running the arc lamp at the local theater.

Several eyewitnesses, including newspaper correspondents, commented that during the descent, the glider had made two complete barrel rolls but immediately righted itself each time. The same flight also inadvertently demonstrated that the revised and strengthened design could handle significant air pressures.[37] When asked later whether the experience was daunting, Wilkie replied, "It demonstrated absolutely that the machine would not collapse if it got out of control by anything jamming; in fact, the accident gave more confidence in the machine, I believe, than if we had had a good flight."[38]

Montgomery was criticized in several area newspapers for the errant Idora Park flight, coming as it had only seven months after Maloney's death. John H. Pierce of the *Oakland Tribune* emerged as John's chief antagonist. Pierce chastised Montgomery, noting the inherent dangers involved with flying machines. Despite the scorn, John and Wilkie conducted further trials at the Montgomery tract located northwest of the intersection of Telegraph Avenue and 41st Street in Oakland. After the glider was repaired, another trial was made. However, as the inflation of the smoke balloon was nearing completion, a weak spot developed in the top, and when a gust of wind came, within seconds the balloon was split from top to bottom.[39]

Montgomery's reputation outside of the Bay Area remained intact. On February 25, an editorial in the *Washington Post* included him in lofty company: "Profs. Montgomery, Langley, and Bell are, of course, eminent scientists, who have unselfishly devoted their lives to the uplifting of other people. They deserve the utmost praise."[40] Sadly, Samuel Langley passed away only two days later, in Aiken, South Carolina, at the age of seventy-one.[41] The same newspapers that had ridiculed Langley for his failed attempts at powered flight now said that he had died of a broken heart, an interpretation shared by both Bell and Chanute.

In early March, Richard Montgomery read in a newspaper article that the U.S. Board of Ordnance and Fortification had purchased gas balloons for aerial observation. They were being stored at the military barracks at Benicia, California, in the northeastern portion of the San Francisco Bay area. Richard wrote to Congressman William R. Knowland, a friend of the

Montgomery family and political representative of Oakland, to propose that these balloons be used to lift the Montgomery aeroplane as a demonstration of its potential for reconnaissance. Richard also proposed that military personnel be trained as glider pilots, thereby relieving John of the necessity for hired pilots. Congressman Knowland forwarded the letter to the War Department, and on April 5 the Board of Ordnance and Fortification endorsed the proposal (with certain conditions) and forwarded it to Brigadier General James Allen of the U.S. Army Signal Corps.[42]

In Allen's reply, he indicated that the Montgomerys were welcome to use the military-owned balloons, but on the condition that all expenses were to be borne by John, including supplying the hydrogen and the pilot and providing pilot training.[43] Furthermore, they required that a full description of the Montgomery aeroplane and flight performance data generated while using the military balloons be provided. Allen responded, "If Professor Montgomery decides to give a demonstration of his aeroplane at Benicia Barracks, provided he can do so without obligation of expense to the United States, a board of Corps Officers will be ordered to witness and report upon the trial."[44] These conditions were unacceptable to John, given that he had neither the finances nor any interest in giving the government the benefit of his scientific knowledge for free.[45] The War Department was still distressed about Langley's failure; the prospect of spending additional federal resources on flying machines was out of the question. As a result, the Montgomerys dropped the entire concept.

On March 17, 1906, the Montgomery "Aeroplane" patent was allowed, with the paperwork signed by Patent Examiner William W. Townsend.[46] In reviewing the specifications for two successful control systems (Montgomery's versus Wright's), the Patent Office had concluded that the respective inventions contained distinguishing, non-overlapping features. At the time, John lacked the funds to pay the filing fees and had to wait a few months to raise enough money. Because of this delay in payment, the Montgomery patent was officially issued some five months after the case was allowed (U.S. Patent No. 831,173). On May 22, 1906, the Wright patent was also allowed (U.S. Patent No. 821,393). In April 1909, the Wrights' patent application for an automatic device used "to provide means for maintaining the lateral control of a flying machine" was rejected by the patent examiner, who cited the patents of Montgomery and Mouillard as prior art.[47] After subsequent revisions, however, it was allowed (U.S. Patent No. 1,075,533).

Having grown weary of the unreliability of balloon-assisted glider launches, in the spring of 1906 John conceived of an entirely new approach for pilot training. This stemmed from a longer-range goal, moving beyond the tandem-wing glider design of 1903–1905 toward another design that he had originally conceived of in 1903.[48] Montgomery's new plan called for a larger flying machine with dual seats so that aeronauts could be given direct instruction during flight. As he recalled in 1910, "to give them the full knowledge in these matters I was formulating plans for a large starting station on the Mount Hamilton Range from which I could launch an aeroplane capable of carrying two, one of my aeronauts and myself, so I could teach them by demonstration."[49] For this larger and heavier glider, a rail-launching apparatus similar to the one used at Otay Valley in 1885 and 1886 was necessary. The steep hills of the Mount Hamilton Range just east of San Jose, offered the appropriate terrain.[50]

By mid-April 1906, John's ambitious new plans were taking form. However, fate would once again intervene. Early one morning in April, all was eerily quiet in the San Francisco Bay region—the foretelling of a singular event that would simultaneously reshape the lives of several million people as well as the history and geographic landscape of central and northern California.

CHAPTER 14

Baptism by Fire

J ust prior to 5 A.M. on April 18, 1906, a magnitude 4.0 earthquake on
the San Andreas Fault quietly rumbled through the Bay Area. Approximately twenty seconds later, a massive temblor hit, its magnitude
roughly 7.8, with an epicenter below the Pacific Ocean, less than two miles
west of San Francisco. For forty-five to sixty seconds, the earth shook vio-
lently throughout central California. The shaking and subsequent fires destroyed more than 28,000 buildings. In San Francisco alone, the earthquake
killed approximately 3,000 and left 225,000 people homeless.[1] The Great
Quake of 1906 ranks among the greatest urban catastrophes in United States
history. It disrupted livelihoods, destroyed property, and forever changed
communities throughout the greater Bay Area.

At the time, John Montgomery's cousin, Archbishop George Montgomery, was presiding over the San Francisco Diocese, which was completely
devastated by the quake. Archbishop Montgomery took a very active role in
the rescue work in San Francisco and in rebuilding the diocese. John's alma
mater, St. Ignatius College, was destroyed, including the most comprehensive ornithological collection west of the Mississippi, the school's science
laboratories, and the beautiful art that had adorned the church. After visit-
ing San Francisco to assist and support the archbishop, John returned to
Santa Clara and wrote a lengthy letter to Chanute, offering his immediate
impressions of the effects of the earthquake:

> The great calamity that has fallen on this portion of our state and es-
> pecially San Francisco has had serious and far reaching effects in every
> sphere of action; and the work on my Aeroplane is no exception. . . . But
> the ruin and destruction of San Francisco is simply beyond conception
> . . . as a consequence there is a great financial depression, in consequence
> of which I have been impelled to discontinue for a time my work with my
> Aeroplane. How soon I shall be able to continue it I cannot say, as the ele-
> ments from which I have been deriving and expected to derive financial
> assistance have been laid waste.[2]

The investors who had pledged to support John in his future plans were now overwhelmed by their losses.[3] Surrounded by chaos and economic hardship, John saw his anticipated financial backing disappear in the rubble and the fires. He quickly lost momentum in the increasingly competitive field of aerial navigation. Richard Montgomery later recalled that after the earthquake, the army was briefly interested in the Montgomery aeroplane: "Then an Army officer came out, saw my brother, and made a proposition for him to take a photographer up with him over San Francisco and take a view [panoramic shot] of the city, which was an impossibility with a one-man machine."[4]

The stress of helping to rebuild the diocese led to months of frail health for Archbishop Montgomery, and on January 10, 1907, he succumbed to complications following surgery.[5] His passing dominated the front pages of newspapers in the Bay Area for two days. The entire San Francisco Diocese attended his grand funeral, including clergymen from all the Catholic orders in the city. "Franciscans, Dominicans, Jesuits, Paulists, Sulpicians and Salesians were there," wrote the *San Francisco Call*, "with the sisters of twelve different orders. Five Bishops joined in the service and gave to the body of the dead the fivefold absolution that is said only for those of high rank in the church."[6]

Thomas Baldwin was also affected by the earthquake, losing five airships worth approximately $8,000 when his large storage facility on Market Street was destroyed.[7] His only surviving airship, *The California Arrow II*, escaped destruction, having been transported to Baldwin's new facility at Hammondsport, New York, just prior to the quake. With financial resources available to him in New York, Baldwin was able to walk away from his losses in California to begin a new phase of aeronautical development with the assistance of Glenn H. Curtiss. It is significant that in this same period, Baldwin and Curtiss tested the thrust generated by the "Baldwin" propeller when it was connected to the Curtiss engine. Curtiss mounted the propulsion system on a three-wheeled "wind wagon."[8] The engine was capable of turning the propeller at a rate of 300 revolutions per minute, which propelled the wind wagon at a speed of 30 to 35 miles per hour.[9] Curtiss's exposure to the Baldwin propeller and his subsequent wind wagon experiments would soon serve him well when he turned his attention to heavier-than-air flight. In fact, in 1908 he would install wings on the front of the "wind wagon" to determine lift and airfoil efficiency and then incorporate the three-wheeled vehicle (with engine, propeller, and wings) as the

Fig. 29. Thomas S. Baldwin and Glenn H. Curtiss with Curtiss's "wind wagon," constructed to test Baldwin's propellers, ca. 1906. Through this interaction, Curtiss learned propeller design. (Courtesy Glenn H. Curtiss Aviation Museum)

platform for the Curtiss-designed *June Bug* and similar powered aircraft designs.

In the months following the earthquake, John redirected his focus toward controlled, laboratory-based experiments on an affordable scale. Wilkie and Defolco were now otherwise engaged, so John hired a mechanic by the name of Cornelius Reinhardt to help with various construction duties in this laboratory. With the assistance of Reinhardt, he constructed elaborate new mechanical devices, including an electrically powered whirling disc and a wind tunnel to further explore and refine airfoil and propeller theory. Model gliders were mounted on the perimeter of a 10-foot-diameter whirling disc to study the lift of different airfoils and combinations of airfoils. With the addition of various propellers fixed to a small electric motor placed on the end of the whirling arm, they could determine which propeller shape provided the most thrust. Reinhardt confirmed that John intended to eventually apply a motor to a Montgomery aeroplane.[10] He assisted John in constructing the new wind tunnel. Constructed of sealed glass, it was 3 feet in diameter and had a motor and fan at one end and a hole with a rod on which to hang the model wings so that he could change the curvature of surfaces from the outside. Reinhardt noted, "The remarkable thing which we always observed was that the flour dust blowing over the front edge of the curved wing surfaces did not merely deflect a little and pass on but always curled upward against the under surface of the wing to lift it."[11]

By 1906, Victor Lougheed's interest in aeronautics had intensified. He spent several weeks with Montgomery at Santa Clara College, where he became familiar with John's work, his aeroplanes, and his experiments in aerodynamics. He participated in a series of wind tunnel tests conducted by John and Reinhardt at the college. The wind tunnel, a slight modification of the one John had built in the late 1880s, was used for measurement of the lift and drift of different specimen surfaces placed within it. It had a glass top that made it useful for observation or for photographing the flow of the airstreams around different objects.[12] Lougheed's subsequent description of the apparatus and associated experiments provide the only surviving account:

> The tunnels used were shallow wooden troughs, eight to ten feet long, and from two to four inches deep and eighteen to twenty-four inches wide. Provided with plate-glass covers through which observations could be conducted, these funnels were used as means of testing the different surfaces from which data was desired. In the form that afforded the most satisfactory results, one end of the tunnel was left open while into the other a blast of air was sent by a centrifugal blower. A series of fine wire-gauze partitions in a canvas double funnel at the end of the tunnel served to render the flow even and to break up eddies and side currents that might originate in the blower. The enlargement served to compensate for the loss of opening due to the wires of the mesh. Also, the use of the canvas between the solid wood of the funnel and the revolving masses of the blower and the electric motor employed to drive it, was found effective in damping out vibrations that otherwise might have interfered with some of the experiments.[13]

The speeds of the air currents ranged up to forty miles an hour. The trial sections, which when thin were bent out of sheet metal and when thick were carved out of wood, were used in two ways. If the objective was simply to observe the course of the air currents, the sections were made long enough to reach from the bottom of the tunnel to the glass cover, which held them lightly in place by resting on them. When lift and drift readings were desired, the trial sections were made short enough that they just cleared the tunnel top and bottom, and were mounted on a light metal arm

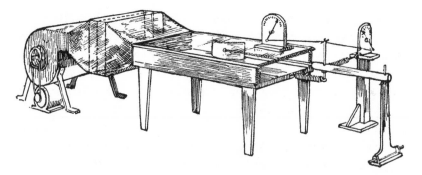

Fig. 30. Montgomery's later wind tunnel. The year of its construction is unknown, but it is probably the one alluded to by Baldwin in 1904. The one above was later used by Montgomery, Reinhardt, and Lougheed in 1907. (Diagram and explanation from Lougheed, *Aeroplane Designing for Amateurs*, 38)

that reached into the end of the tunnel and was pivoted such that the surface under trial could move freely in any direction—as well as be set in any position.[14] In this period, John gave Lougheed demonstrations of *The Defolco* at Villa Maria through unmanned ballasted flights from the derrick with the assistance of Reinhardt. In France during the late summer and fall of 1906, Alberto Santos-Dumont embarked upon a new path that would prove to be influential in heavier-than-air powered aeronautics. Although some have argued that the Wright Brothers' flights at Huffman Prairie in 1904 were made out in the open and within sight of the nearby streetcar line, they specifically timed their attempts to occur between scheduled trains. Those flights did little to convince the public that they had actually flown. However, after some preliminary and modest flights in September and again in October, on November 12 Santos-Dumont electrified the world with his aircraft *14-bis*. He made several flights of up to 722 feet at Bagatelle, France, in front of a large audience, including official observers from the Aero Club of France. His charm and generosity in providing his technology to the world earned him tremendous popularity. Once again, as with his earlier dirigibles, Santos-Dumont captured the imagination of the public with his aerial exploits. He was soon hailed as the first man to make a public flight with a heavier-than-air powered flying machine. In September 1906, Danish inventor J. C. H. Ellehammer also made successful public flights.

During October 1907, Montgomery traveled to New York City to attend the International Aeronautical Congress, scheduled for the last week of October. The congress was once again organized by Albert Zahm, and attracted many people who were interested in aeronautics, including Octave Chanute. At the meeting, John presented a theoretical dissertation titled "Principles Involved in the Formation of Wing Surfaces and the Phenomenon of Soaring." The material was essentially a combined summary of his unpublished 1894 thesis "Soaring Flight," his 1897 lecture to the Southern California Academy of Sciences, and material excerpted from *The Redwood* and *The Aeroplane* in May 1905 and from the November 22, 1905, *Scientific American Supplement*.[15] John's lecture at the congress was published in the *Aeronautics* issues of October through December 1908.[16] Although some authors have suggested otherwise, there is no indication that his lecture or the subsequently published material was met with a negative reception from Chanute or others in attendance.

By 1907, interest in heavier-than-air flight had been rekindled in northern California. A new generation of enthusiasts now singled out John Montgomery for his leadership and expertise. Under John's instruction, on December 23, 1907, San Franciscan Cleve T. Shaffer, an inventor and emerging aviation enthusiast, made his first solo flight in a glider of his own construction.[17] Shaffer later distinguished himself as the founder of the California Aero Manufacturing and Supply Company, and became an influential promoter of aeronautics on the West Coast. Cleve's sister, Geneve, also took an active interest in aeronautics and assisted her brother with glider construction. She would go on to become a well-respected aeronaut in hot-air balloons, winning the Portola Balloon Race at San Francisco on October 31, 1909.[18]

At the start of 1908, Alexander Graham Bell's group on the East Coast (Glenn H. Curtiss, Thomas E. Selfridge, Frederick W. "Casey" Baldwin, and John A. D. McCurdy), known as the Aerial Experiment Association (AEA), commenced work on their own flying machine. They decided to build a glider first, to gain practical experience in flight. In January, Chanute provided Bell with plans for a Chanute-type biplane glider. Although not a member of the AEA, Thomas Baldwin was still associated with Curtiss at this time. Just as Curtiss had benefited from Baldwin's expertise in propeller theory in 1906–1907, now he and his AEA associates benefited from Baldwin's expertise as a former "student" of Montgomery's in aeroplane construction. The resulting aircraft, known as *The Hammondsport Glider,* was tested in late January, and AEA members soon became proficient with it at short flights at Hammondsport.[19]

Over time, the U.S. Army came to the realization that navigable flying machines such as dirigibles and aeroplanes could make an important contribution to the nation's security. Brigadier General James Allen, the army's chief signal officer, established an Aeronautical Division as part of the Signal Corps on August 1, 1907, making it responsible for "all matters pertaining to military ballooning, air machines, and all kindred subjects."[20] The War Department established a competition for contracts in both heavier-than-air and lighter-than-air machines. Baldwin and the Wrights quickly seized the opportunity. After an initial meeting with members of the Board of Ordnance and Fortification in early December, Wilbur Wright determined that the War Department had been allotted a maximum of $25,000 to spend on a heavier-than-air flying machine. Shortly thereafter, a tentative specification for one was forwarded to the Wrights, including blank spaces where they could insert their own specifications for the design and performance characteristics, including minimum airspeed, minimum distance flown, and minimum carrying capacity. The specifications they indicated essentially matched the capabilities of the Wright aircraft, and were well beyond what any potential competitor could be expected to produce.[21]

In the late spring of 1908, Baldwin submitted a new dirigible for the U.S. Army contract. The specifications required that it had to be maneuverable in all directions. Baldwin's airship was the best he had to offer. The 20,000-cubic-foot dirigible was powered by a 20-horsepower Curtiss water-cooled engine, which, when actuated through a shaft transmission and tractor propeller, generated a cruising speed for the ship of 20 miles per hour. Counter-rotating twin propellers were mounted on the single shaft for additional propulsion.[22] The pilot sat behind the engine, with biplane stabilizing and trimming surfaces mounted at the aft of the aircraft, while the stern held a cruciform tail for vertical and directional control.[23] Baldwin and Curtiss conducted the trials at Fort Myer, Virginia. On the basis of its success, on August 28 the Baldwin airship was formally accepted by the Aeronautics Board and was designated as *Signal Corps Dirigible No. 1* (SC-1).[24] It was the U.S. Army's first powered aircraft. Baldwin and Curtiss then trained Lieutenants Frank P. Lahm, Thomas Selfridge, and Benjamin D. Foulois to pilot the first military dirigible. Because of Baldwin's important connection with the embryonic U.S. Signal Corps' Aeronautics Division, when World War I broke out he was assigned to be the chief inspector of the U.S. Army's airships (dirigibles) during the war, with the rank of major. The self-appointed "Captain" had finally received an official military title. As late as July 1908, Baldwin was still claiming publicly that he had received scientific training from John Montgomery at Santa Clara College, to which even he now attributed his successful development of propel-

Fig. 31. Lincoln Beachey (background) and A. Roy Knabenshue (foreground) in an airship race at St. Louis, October 1907. Baldwin's *California Arrow* widely influenced airship design in America during the period 1904 to 1910. (From Francis T. Miller, *The World in the Air: The Story of Flying in Pictures* [New York: G. P. Putnam, 1930], 2:122)

lers in the highly influential *California Arrow.*[25] *The Arrow* would remain the basic platform for most American dirigible enthusiasts through 1910.

Only seven months later, on July 4, 1908, Glenn Curtiss made the first public flight in America of a powered heavier-than-air flying machine at Hammondsport, capturing the *Scientific American* Trophy. (There is some disagreement in the literature about whether Casey Baldwin's March 1908 flight in the AEA *Red Wing* qualified as the first "public" powered flight, given that it was a private flight trial.) Soon thereafter, the Wrights conducted their trials for the U.S. Army Signal Corps at Fort Myer. After delays caused by a tragic accident (which caused the death of Lieutenant Thomas Selfridge and an injury to Orville), Orville completed the required flight performances in July 1909.[26] Having finally secured contracts that satisfied their terms, the Wrights now agreed to "go public" by making a series of flights (Wilbur in France and Orville at Fort Myer) to prove the performance of their product. In the process they generated widespread interest and earned tremendous fame, and they soon consolidated their financial interests. They made additional contracts for the manufacture and licensing of Wright flying machines in Europe.

Chanute became increasingly discouraged by what he viewed as the Wrights' overemphasis on money, the exorbitant prices they charged for their product, and their inability to acknowledge the accomplishments of their predecessors. As he wrote to fellow experimenter George A. Spratt, "I think myself that they have unduly ignored the assistance which has been rendered them by others, but that is natural for men who want to make a fortune."[27] Nevertheless, there is no doubt that the demonstrations by the Wrights, Curtiss (in particular his dramatic win at an aviation meet in Reims, France, in August 1909), and European aviators ignited a public fascination with flight. It was only at this time that John Montgomery, along with the majority of Americans, became aware of the Wrights' achievements.

Still attempting to find additional sources of revenue, John focused his attention on wireless telegraphy during much of 1908. Telegraphy had been the subject of a great deal of interest and innovation since the turn of the century. A group of San Francisco businessmen, headed by local politician and entrepreneur Christopher Augustine Buckley, joined forces to exploit developments such as telegraphic printing. They formed a company called the United States Wireless Printing Company (USWPC), which was capitalized at $10 million, and opened a factory in San Francisco. One of the directors of the company, Dandridge H. Bibb, would become an important figure in John's career in electricity. Buckley and his associates scouted for inventors who were capable of producing new technology in telegraphy that could be exploited for business development. They soon became aware of John's reputation as an expert in electricity.

In the summer of 1908, Buckley's group established the Burlingame Telegraphic Typewriting Company in Los Angeles, using much of the factory equipment from the now-defunct USWPC facility. An agreement was reached with John, and he moved to the Burlingame facilities for a few weeks. There he developed a system known as a "teleautoprint," consisting of a telegraph connected to an electric typewriter. The device was capable of two-way message transmission through a relay system using the typewriter as the keyboard. The receiving instrument automatically printed incoming messages as they were received. By regulating both transmission and reception, the teleautoprint prevented the sort of "wire-tapping" that was problematic with the use of standard telegraphy. On the strength of John's invention, Buckley and his investors formed a new company known as the Teleautoprint Company of San Francisco. When they learned that a similar device had been announced on the East Coast, however, they discontinued their financial support of John's work.

John also revised and improved the astronomical observatory at Santa Clara College. He developed an electric "rectifier" to convert alternating electric current into direct current that supplied power to various scientific instruments and then returned the current to a powerful three-cell storage battery. This efficient system provided the energy required for simultaneous measurements of astronomical and meteorological data, a task that had previously been possible only for very short periods.[28] In 1908, he also helped install the school's first seismograph (a Wiechert 80 kilogram Enz model) to measure the energy released in earthquakes.[29]

Late in 1908, John wrote an account of his flight experiments up to 1906. Titled "Some Early Gliding Experiments in America," the article was published in the January 1909 issue of *Aeronautics*. He penned another article, including attached sketches, around the same time (late 1908 to early 1909), briefly summarizing the evolution of his flying machines from 1884 through the 1903–1905 tandem-wing model. The original title, "My Experiments in Flying," was changed to "The Early Development of the Warping Principle" sometime later (most likely 1908–1909), but the article was never published. Given Montgomery's earlier article on ornithology ("Our Tutors in the Art of Flying") and his lengthy paper on aerodynamics ("Principles Involved in the Formation of Wing Surfaces and the Phenomenon of Soaring"), it appears that in 1907 and 1908 he was finally following Chanute's advice by summarizing his theories and practical experiments so as to attract potential supporters interested in financing further research and development in aeronautics.[30]

As the sport and science of aeronautics continued its dynamic course in America, the Aero Club of America remained in a state of transition because members' interest was divided between lighter-than-air and heavier-than-air machines. In June 1908, a faction of the club had become so frustrated by the conflict that they formed their own group. The Aeronautic Society of New York (ASNY) met for the first time in July 1908. Among the forty founding members, the more prominent were William J. Hammer, Hudson Maxim, A. Leo Stevens, Hugo C. Gibson, and Ernest L. Jones, with French pilot Henri Farman as an honorary member. The ASNY was, in essence, precisely the type of organization that Chanute had wanted to establish after the 1893 Aerial Conference at Chicago. It was the first aeronautical society in the country to provide a flying site for its members (at Morris Park, New Jersey) and the first organization other than the U.S. government to purchase a heavier-than-air flying machine, the Curtiss *Golden Flyer*, for which it paid $5,000. The society provided workshops on aviation, offered classes in aeronautics, and sponsored contests and prizes to stimulate innovation. Montgomery's work in aviation was known to its members, some of whom had met John at the International Aeronautical Conference at New York in October 1907.

Inspired by public demonstrations of powered aircraft throughout 1908 and 1909, a new wave of interest in aviation swept through America. Californians began experimenting with their own heavier-than-air designs. The Aero Club of California was formed in Los Angeles on March 26, 1908. Some thirteen months later, on May 1–2, 1909, the club sponsored the first aviation show in California, featuring gliders made by Edgar S. Smith and W. J. Cochrane, a full-scale model of a J. H. Klassen helicopter called *The Gyroplane*, and an airship designed by A. L. Smith. Prominent experimenters and enthusiasts in southern California also included Harry La Verne Twining (a physics instructor at Los Angeles Polytechnic High School and California correspondent for *Aeronautics*), J. S. Zerbe (an aviation enthusiast and author), Thaddeus S. Lowe (a pioneer balloonist), Charles H. Lamson (a friend of Chanute's and a flying machine inventor who later filed a lawsuit against the Wrights in January of 1910), and in particular A. Roy Knabenshue, who continued to be a leading figure in aeronautics based in Los Angeles during 1906–12. In addition, Glenn L. Martin built an aeroplane in July 1908 that was powered by a Ford engine and successfully flew it 100 feet on August 1, 1909. This has traditionally been considered the first powered flight of a heavier-than-air flying machine in California.

Other prominent California-based experimenters included Bay Area glider manufacturers Cleve T. Shaffer and Jack T. Whitaker. Spurred by the formation of the Aero Club of California in Los Angeles, in March 1909 Shaffer and Whitaker founded the Pacific Aero Club at San Francisco so that aircraft builders could share information and pool their resources.[31] Montgomery was invited to be a founding member. Shaffer worked closely with his younger sister Geneve in the design and construction of a biplane glider, which they flew from San Bruno Mountain south of San Francisco. Geneve soloed in that glider on August 1, 1909.[32] For this achievement, she is credited by the Smithsonian Institution as the first woman glider pilot in the United States, although there were no glider licenses awarded at this time.[33]

She also worked with Cleve at the Shaffer Aero Manufacturing and Supply Company, serving as secretary and chief rigger. At this time the Oakland Aero Club was formed, with Park Van Tassel and Albert Van der Naillen as prominent members and Richard Montgomery as a member. Other notable California enthusiasts during this period included George H. Loose of Palo Alto, a Pacific Aero Club member who designed and constructed a series of monoplanes, including one that flew in 1909; Fred Wiseman (auto racer and aviation enthusiast); Oakland natives Peter and W. P. English, designers of a helicopter during 1906–1908; Weldon B. Cooke of Oakland, whose training as an aviator began with a Montgomery-type glider; and the enigmatic Lyman Gilmore, Jr., of Nevada County, who began his initial experiments with a powered aircraft in August that was allegedly based on a design he had developed in 1898. Also notable are Donald H. Gordon and Waldo Dean Waterman, who flew biplane gliders near El Cajon and San Diego, respectively. The California Balloon Club was also formed, with members from both the Pacific Aero Club and the California Aero Club, most prominently Thaddeus Lowe, Dick Ferris, and A. Roy Knabenshue.

In 1909, international recognition of Montgomery was furthered through a study conducted by the Austrian Technical Flying Society and the Austrian Imperial Army to determine who truly had been "first to fly" a heavier-than-air machine in controlled flight.[34] The Technical Flying Society was patronized by leading scientists in Germany and Austria. The results of their study were presented through the Vienna-based *Fachzeitung für Automobilismus und Flugtechnik,* the society's official journal. After their search, they concluded that John J. Montgomery had been the first in the world to make successful controlled flights in a heavier-than-air flying machine. Their official letter, quoted in a leading newspaper, stated in part: "If any person be entitled to the glory of having performed the greatest flights of gliding it is not Lilienthal, but Montgomery. . . . and it has taken Europe until now to find out that Professor John J. Montgomery made successful flights in 1884, ten years before Lilienthal's record."[35]

Throughout this same time, John continued his passion in electrical engineering and completed the development of a new invention, an alternating electric rectifier. A practical application for the rectifier was to charge batteries using the low-voltage currents normally used in residences. It was particularly useful for electric automobiles and boats but could be used for any electrical application that used batteries. End users could charge an automobile battery overnight without resorting to standard rectifiers, which were priced at $200 each. On August 28, 1909, Montgomery applied for U.S. patents for his electrical rectifier and a "process for compelling electric motors to keep in step with the waves or impulses of the current driving them."[36]

As Victor Lougheed's interest in aviation continued to grow, he began collecting material on aviation progress and sought the counsel of prominent men in aviation. At the time of his visits with Montgomery in 1907, he asked John to provide him a letter of introduction so that he could meet Octave Chanute. Chanute and Lougheed subsequently became friends and corresponded frequently until Chanute's death three years later. Lougheed later confirmed that his expertise in aeronautics was based substantially on interactions with both Montgomery and Chanute between 1907 and 1910. Lougheed held many patents, but in particular he sought patent protection for aeronautical devices such as airfoils and propellers.[37] It is noteworthy that Lougheed's 1919 patent application for improvements to an airfoil specified that it was for "an improvement based on the Montgomery invention."[38] In 1911, Victor designed a biplane for his half-brothers that Allan later flew over a period of several years, eventually accumulating 35,000 miles.[39] Victor's expertise would be crucial in nurturing Malcolm and Allan's technical expertise as aircraft designers. As a result of this interaction, Allan and Malcolm would become highly successful and influential aircraft designers and manufacturers. Allan and Malcolm initially formed the Alco Hydro Aeroplane Company in San Francisco in 1912 and then relocated to Santa Barbara in 1916, where the brothers formed the Loughead Aircraft Company, which survived until 1926. Raymond Acre was a silent partner in this venture. At that time Allan relocated to Hollywood and formed the Lockheed Aircraft Company, which became a cornerstone of the American aircraft industry.

In 1908 and 1909, Lougheed prepared an exhaustive compendium on aviation. He provided Chanute with an advance manuscript of the book for his review and comment.[40] *Vehicles of the Air* presented Montgomery's work in greater detail and more prominently than any prior work. It was released in November 1909 by American, British, and German publishing houses and became a standard reference on the subject, ultimately selling more than 30,000 copies worldwide over three editions in 1909, 1910, and 1911.[41]

In the summer of 1909 the sale of the Glenn Curtiss *Golden Flyer* to the Aeronautic Society of New York was indirectly responsible for a new impetus in American aeronautics. The Wrights responded immediately, filing a lawsuit in August against the ASNY, the Herring-Curtiss Company, and Curtiss individually, alleging infringement on their aeroplane patent. Following an exhibition in St. Louis on October 12 and 13, Curtiss visited Chanute in Chicago to review the lawsuit. He came away convinced of his legal position and happy to have Chanute as an ally. Following the meeting, Chanute wrote to Curtiss's attorney, Emerson Newell, confirming his opinion that "the bare idea of warping and twisting the wings is old, but there are several ways of accomplishing it."[42] Curtiss subsequently sought the advice of Alexander Graham Bell and John Montgomery. Montgomery held a private meeting with Curtiss and a patent attorney, Raymond I. Blakeslee, in November.[43] Rumors circulated in the Bay Area that he was to provide a deposition in the lawsuit between the Wrights and the ASNY.

At this same time, Lougheed and a young automotive engineer named Raymond Acre convinced James Plew, an agent for the White Automobile Company in Chicago, to get into the aviation business. Plew was a former railroad conductor who had gained his wealth from the invention of a bicycle seat. He purchased a Curtiss-powered aircraft, but Acre was able to make only brief hops just a few feet off the ground. A larger engine was installed, and Jimmy Ward and Otto Brodie, two young aviators who were associated with Lougheed and Plew, flew the plane at exhibitions in Chicago. Lougheed envisioned greater ambitions for aviation. After holding preliminary talks in November at Santa Clara College, he and John formed a business agreement in early 1910.[44] Under the Montgomery-Lougheed contract, Lougheed and his partners would manufacture and sell powered aircraft based on a Montgomery design. Lougheed would be obligated to handle any lawsuits that might arise from perceived patent infringement. John absolved himself of any business responsibilities and focused all of his attention on the research and development of the aeroplane.

In November 1909, the Wright Brothers joined forces with Wall Street capitalists J. P. Morgan and Charles R. Flint and incorporated the Wright Company for the manufacture of aeroplanes and to protect the Wrights' patents in the U.S. and Canada. The investors included Cornelius Vanderbilt, August Belmont, Robert J. Collier, Howard Gould, Andrew Freeman, and Russell A. Alger, powerful men who were associated with large companies in transportation, finance, and mining.[45] The Wright Company opened with $1 million in capital. Secure with this financial backing, the Wrights launched an aggressive and comprehensive public relations and legal campaign to prevent others from profiting from the exhibition or manufacture of heavier-than-air flying machines and to collect royalties from past profits made by others in the manufacture, selling, or exhibition of such craft.[46] In an era when the public generally viewed monopolies with contempt, the Wrights' decision to ally themselves with powerful capitalists did not help win sympathy for their cause.

John viewed the Wrights' overall position as deeply irrational. He felt compelled to revisit his priority in heavier-than-air flight. Montgomery's reaction to the campaigns initiated by the Wrights in 1909 would forever change the way he was represented in Wright-themed publications.

CHAPTER 15

Birds of Prey

I am certainly not going to stand back and allow a check to be put on man's right to fly.

John J. Montgomery, 1910

In the winter of 1909–10, final preparations were completed for the first International Aviation Meet in America, to take place January 10–20, 1910, at Dominguez Hills near Los Angeles. The success of the event hinged on the attendance of Glenn Curtiss as well as Louis Paulhan, France's leading aviator. Paulhan had received his initial flight training in the fall of 1904 in a tandem-wing glider that he had constructed with Louis Peyret. Fearing that Curtiss and Paulhan's participation in the meet would damage the Wrights' commercial interests, the brothers moved to derail the event, firing something of a warning shot to intimidate others who might attempt to make money in aviation.

On January 3, the Wrights obtained a temporary injunction against both the Herring-Curtiss Company and Glenn Curtiss himself regarding the manufacture, exhibition, and sale of aeroplanes. On that same day, Paulhan arrived in New York City, bringing with him two *Blériot XI* aeroplanes and one *Farman* aeroplane. These were the most advanced aeroplanes in Europe, and perhaps in the world. Aware of the Wrights' legal moves, Blériot had altered the control mechanism on his aircraft so as not to be challenged by a lawsuit. Changing the angle of incidence on the wing was accomplished by a stick held in the left hand, while the rudder was controlled by a pedal. According to Blériot, "These two effects are completely independent of and in no way corrective [as in the case of the Wrights' use of their rudder]."[1]

Shortly after his arrival in New York, Paulhan was presented with a federal summons issued by Judge Billings Learned Hand, barring him from taking part in exhibitions with the *Farman* aeroplane. A bond was delivered, and Paulhan proceeded with his plans; he had been advised by his attorneys in New York that his control system was sufficiently different from the Wrights' patent claims, but they were not certain about the eventual outcome of the injunction. One author during the period noted Judge Hand's reason for approving the injunction: "The use of the ailerons may be far more casual and unusual than the continual bird-like warping of the Wrights' wings, but, thought the court, it is the principle itself which is involved and not the frequency of its use or its mode of appliance."[2] This broad interpretation was starkly different from the opinions of some highly knowledgeable aeronautical inventors, including Chanute, Alexander Graham Bell, and Hiram Maxim.

The International Aviation Meet at Los Angeles was a huge success. Curtiss, Paulhan, and other aviators provided the West Coast with its first large-scale exhibition of powered heavier-than-air flight. Paulhan proved his stature as a world-class aviator and charmed the press. Records were broken for altitude and airspeed. Because of the injunction, Curtiss had to fly his aircraft using only the elevator and ailerons for control. In addition to heavier-than-air craft, Knabenshue, Beachey, and others made flights with dirigibles based on *The California Arrow*, which continued to serve as a template for airships in America.

Immediately after the meet, northern California hosted its own Second International Air Meet on January 25 and 26 at the Tanforan Racetrack, just south of San Francisco. The event was organized by the Pacific Aero Club through the leadership of Cleve Shaffer. Montgomery attended this meet, where he witnessed powered heavier-than-air flight for the first time. Yet again, Louis Paulhan made an impressive showing, in spite of adverse weather. Montgomery and Paulhan met several times to discuss aeronautics, and in particular John's theories on soaring flight, which Paulhan found fascinating.[3]

In late 1909, members of the Aeronautic Society of New York encouraged Montgomery to relocate to New York and renew his aviation experiments at their facilities. Given his leadership and his pioneer status, John would be an important ally for the besieged ASNY. On January 21, 1910, patent attorney Raymond I. Blakeslee met once again with John to review the Wright-Curtiss and Wright-ASNY lawsuits. Although John had recently implicated Curtiss in his public statements regarding potential infringers, he agreed to meet with Blakeslee to discuss matters more fully. Following the meeting, local newspapers suggested that Montgomery had generated affidavits for presentation in the lawsuit.[4] Both the *Oakland Tribune* and the *San Francisco Examiner* noted that the ASNY had appealed to Montgomery to testify in the upcoming lawsuit.[5] Instead of relocating to New York, he chose to stay in California.

The Wrights' legal and public relations strategy was to present their contributions to aviation in a very comprehensive manner, in contrast with their predecessors, who were usually characterized as "inspirational" but generally unsuccessful. Accordingly, their interviews and writings regarding prior art were filled with broad generalizations and at times misleading assertions. In one court deposition, for example, Wilbur Wright summarized the contributions of three generations of inventors in aviation as "merely a little experimenting and much speculation."[6] In writing about the Wright patent litigation, one author of the period noted, "The decision of the court reflects the popular passions of the moment, the shrewdness of a lawyer capitalizing on the weakness or confusion of a given witness, and the judge's limited ability to understand the technical points involved."[7] History is replete with examples of technological breakthroughs that evolved through a process that includes parallel invention, adoption, adaptation, and refinement. The Wrights' promotions, however, countered the actual process of how invention occurs in a basic discipline such as aviation. Subsequent to their active pursuit of financial reward, the Wrights began using their unique position to bias public perception and history.

Many experimenters and inventors who worked in aviation between the 1880s and the early 1900s viewed the pursuit of flight as something that all humanity should have the right to engage in freely. John Montgomery was among those who expressed incredulity about the Wrights' stance. From November 1909 well into 1910, John gave a series of interviews to clarify his position on the matter for the public, establishing himself as a clear adversary of the Wrights. His statements in an article in the *New York American* titled "Aeroplane Father Denounces Wrights" noted the absurdity of the Wrights' claims to be *the* sole originators of the airplane. As John pointed out in the interview, "Professor Langley, Dr. Chanute, Professor Zahm, and I contributed to the world the fundamental principles that have made the aeroplane a success." He added, "The right to fly should belong to every man. . . . Aerial highways should be as free to men as they are to the birds. If Professor Langley could speak to-day I am sure he would be severe in his criticism of the Wrights."[8] Hiram Maxim, an American living in the United Kingdom who was a prolific inventor with extensive experience with aviation and with the American and British patent processes, asserted, "I have no doubt, however, that a determined effort on the part of American aviators, if supported by money, would be quite able to greatly curtail the preposterous claims of the Wrights."[9]

With the formation of such diametrically opposed camps, Chanute was left in an awkward position. The Wright-Chanute correspondence of 1900–1905 reveals the comprehensive nature of his role in the Wrights' progress in aeronautics. Yet by the time of their legal campaigns, the Wrights had begun to describe Chanute as having provided little more than encouragement. Chanute lamented in a letter to George Spratt that Wilbur Wright "has greatly changed his attitude within the past three years. . . . The Wrights used the conceptions of others, made them practicable and deserve the reward for those successful efforts."[10] In another interview of the period, he stated:

> Now as to the wing warping patent, under which they demanded damages from others for alleged infringement, this is not an absolutely original idea with them. On the contrary, many inventors have worked on it, more or less successfully, from the time of Leonardo da Vinci. Two or three have actually accomplished short glides with the basic warping idea embodied in their machines.[11]

Chanute's personal correspondence with the Wrights shows that there were deeper divisions than were revealed publicly. After one of his interviews was published in the *New York World* on January 17, 1910, Wilbur wrote to challenge him. Chanute replied:

> I did tell you in 1901 that the *mechanism* by which your surfaces were warped was original with yourselves. This I adhere to, but it does not follow that it covers the general principle of warping or twisting of wings, the proposals for doing this being ancient. . . . Therefore it was that I told you in New York that you were making a mistake by abstaining from prize winning contests while public curiosity is yet so keen, and by bringing suits to prevent others from doing so. This is still my opinion and I am afraid, my friend, that your usually sound judgment has been warped by the desire for great wealth.[12]

Wilbur responded with further arguments and expressed his desire to put the disagreement behind them, but it was too late. The breach between the Wrights and Chanute, which had began to appear as early as 1903, was now irreparable.

The issue of whether or not the aileron itself was an appropriation of the Wrights' intellectual property was raised once again when the U.S. Patent Office approved a patent by the AEA covering, among other things, "lateral balancing rudders" and their control. The term "rudder" in this instance was used in lieu of the more common term "aileron." Alexander Graham Bell, then a regent of the Smithsonian Institution, publicly avoided the Wright litigation matter. However, he did say that successful heavier-than-air flying machine technology was not the exclusive property of the Wrights, nor was the principle of varying the angle of incidence on the wings to maintain lateral balance. "We believe that the Wrights should have all that they are actually entitled to," he remarked, "but I have never believed that their rights conflicted with our devices."[13]

In late March 1910, Lougheed and Plew formed the Aero Club of Illinois. Montgomery was elected as an honorary member. During the second week of January 1910, just before the San Francisco International Aviation Meet at Tanforan, Montgomery had sent *The Defolco* to Lougheed in Chicago, so that Lougheed and his associates Raymond Acre and Edmund F. Andrews could build and test replicas. Andrews, a young man who hailed from a wealthy Chicago family, had been befriended by Acre because of their shared interest in aviation. As part of their business arrangement, Lougheed was to develop an engine for a new Montgomery monoplane designed by John in California. Plew would have the manufacturing rights to any Montgomery monoplanes resulting from this plan. Lougheed later related that when Chanute heard in March that Lougheed, Acre, and Andrews were assembling *The Defolco* at the White Automobile Company factory in Chicago, he was eager to see the glider and stopped by to view it. The April issue of *Aircraft* noted that Chicago's first aeroplane factory was preparing to open.[14]

Victor Lougheed and his associates constructed a replica of *The Defolco* (hereafter referred to as *The Lougheed-Acre Replica*) and tested it to determine its gliding angle and how much motive power would be required to launch it into the air.[15] Lougheed and Acre conducted gliding flights from the large sand dunes along the southeastern shore of Lake Michigan. This same region had been used for gliding fourteen years earlier by Chanute and his associates. They also tested the glider in towed flights behind a yacht belonging to the Andrews family (*The Lillian Second*) on Lake Michigan with a 10-mile-per-hour headwind. The glider was ballasted with sandbags at the pilot's position (the controls were clamped in a fixed position) and towed with a 75-foot-long tow rope incorporating a Fairbanks spring balance to measure drift.[16] "It had the lowest angle of gliding I had encountered," Lougheed concluded.[17]

Fig. 32. Victor Lougheed's powered replica of *The Defolco*, constructed in the late spring of 1910 in Chicago. (From Lougheed, *Vehicles of the Air*, 435)

In 1910, Lougheed increased the glider's wing area, modified the frame, and installed a 7-to-8-horsepower gasoline engine developed by Carl S. Bates of Chicago. The resulting aircraft had 250 square feet of wing area, a gross weight of between 200 and 250 pounds including fuel and pilot, and a wheeled undercarriage. This *Lougheed-Acre Replica* was kept in an "aviation camp" on the south side of the Chicago railroad yards where Acre, Andrews, Bates, and other aviation enthusiasts, including William Avery, Paul Butusov, and Horace B. Wild, stored their heavier-than-air flying machines. Malcolm and Allan Loughead were daily visitors to the aviation camp.

During this time, Acre attempted a powered flight with *The Lougheed-Acre Replica*. Like the Curtiss pusher, it skimmed over the ground at a high rate of speed without taking to the air because of an underpowered motor. However, the larger goal of generating a fully functional powered flying replica of the Montgomery aeroplane suffered from a lack of persistence. Plew's business prospects in aviation were limited to a few public exhibitions with the Curtiss pusher piloted by Ward and Brodie.

John traveled to New York to deliver an invited lecture to the ASNY on April 21. In light of the recent controversy, he described the details of his own aeronautical development from 1884 through 1905. In the course of his talk, he noted that this lecture was a reprise of his second lecture at the Chicago Conference on Aerial Navigation in 1893. He avoided any reference to the Wrights, preferring to keep such comments outside of scientific venues. That evening, the membership of the ASNY voted unanimously to give Montgomery honorary membership in the society, "in recognition of his manifold labors to advance the art of aviation."[18] This important speech was subsequently excerpted in the *New York World,* the *New York Sun,* the *San Francisco Examiner,* the British serial *Aeronautics* (May 1910), the American serial *Aeronautics* (November 1911), and then (paraphrased extensively) posthumously in Charles B. Hayward's *Practical Aeronautics* (1912) and printed in full in the *Engineers' List* (January 1912).[19] It should be noted that Orville Wright wrote the introduction to "Practical Aeronautics" and was responsible for the footnoted objection (p. 159) to Montgomery's claim of an early use of wing warping, but he passed on the opportunity to challenge Montgomery's descriptions of successful gliding flights and technology with his early series of gliders in the period 1884–86 presented on the same page. In the summer of 1910, British aeronautical authority Baden Baden-Powell published an editorial in the London *Aeronautics* noting Montgomery's ASNY lecture, and placing John in the context of the Wrights' pursuit of other aviators at that time:

> Professor Montgomery's lecture certainly makes it clear that he himself adopted this device [warping] in his experiments made in 1884 and 1885, which were detailed before the Aeronautical Congress held at Chicago in 1893. . . . It may, at any rate, serve to settle a controversy that gives signs of becoming very acute. . . . On the face of it, Professor Montgomery seems to have forestalled the Wrights most effectively, while the priority of the other inventions appears to be clearly established.[20]

Although John's public challenges had no appreciable effect on the Wrights' patent litigation, as a result of his series of interviews and talks, the aviation community was becoming increasingly aware that his successful aeronautical technology dated back well before the turn of the century. In April 1910, the ASNY membership learned that their society president, Cortlandt Field Bishop, was in direct negotiations with Wilbur Wright to determine the financial conditions under which the Wright Company would allow the Aero Club of America to sanction international aviation meets in the United States. The agreement prevented promoters from holding such meets without first obtaining a license from the Wrights, who would receive a fee. As a result of this negotiation, a rift formed within the Aero Club, and in May the two groups officially broke away and formed the Aeronautic Federation of America and the American Aeronautical Association, respectively.[21] They later joined forces to form their own national governing body, the American Aeronautic Federation (AAF).

Fifty-five clubs and societies were involved in the movement to form a new national organization to govern aeronautics separate from the Wrights' interest. The membership included, among others, Thomas A. Hill, James Plew, and Victor Lougheed. This freed members from the previous governing body, the Aero Club of America (the United States' representative to the Fédération Aéronautique Internationale [FAI]), and allowed AAF members to avoid the Aero Club's oppressive agreement with the Wrights. John was elected a member of the AAF and served on the committee (representing the ASNY) for the inaugural convention scheduled for late June 1910.[22]

In early 1910, John demonstrated his electric rectifier to San Francisco-based venture capitalist Charles S. Wheeler and his associates. By March he had signed a contract for the manufacturing rights and received an advance payment of $2,500 ($58,000 in today's dollars) out of a promised total payment of $10,000.[23] After securing the rights to the invention, Charles S. Wheeler, Dandridge H. Bibb, and a third investor, E. N. Winslow, formed the Wheeler Keepstep Motor Company.[24]

With this unprecedented financial security, John was finally in a position to reexamine other aspects of his life. The romance that had briefly emerged between him and Regina Cleary in 1893 was reawakened. This time, however, John was prepared to start a family. Regina, now thirty-seven and an accomplished pianist, was still smitten. Following a short courtship, they were married on June 30, 1910, at San Francisco's landmark St. Mary's Cathedral. Through a rare courtesy extended by the diocese, the cathedral was closed to the public during the private wedding ceremony. Although fifty-one years of age, John was the first of his siblings to marry.

John and Regina put their honeymoon on hold so that John could attend to some business obligations. They traveled to Chicago in July, where he met with Lougheed and Acre to discuss the progress of and future plans for their aeroplane project. He also attended the inaugural convention of the American Aeronautic Federation. The youthful Regina encouraged John to continue his pursuit of aviation and helped him gain a renewed sense of confidence. John concluded that the time had come for a new series of experiments with a new glider designed at Santa Clara. Acre confirmed that at the time of Montgomery's visit to Chicago in July 1910, plans were already under way for the development of Montgomery's new monoplane.[25]

Although the news of John's invention of an electric rectifier had circulated in the national press in August and September 1909, when word

Fig. 33. Regina C. Montgomery, ca. 1910. Regina was a talented pianist and artist, as well as a published poet. (Courtesy Santa Clara University Archives and Special Collections)

spread in the spring and summer of 1910 that the rights to his device had been sold, Christopher Buckley and his partners brought suit against Montgomery and the Wheeler Keepstep Motor Company. They claimed that the rectifier had been invented during his employment with the Burlingame Telegraph Typewriting Company in Los Angeles.[26] Although the allegation was untrue, John was forced to defend himself through a protracted and expensive legal battle with Buckley that extended through May 1911. This court case prevented him from making serious progress with his new monoplane. While the legal battle ensued, on November 1, 1910, John was awarded patents for the rectifier (U.S. Patent No. 974,171) and for the process used with this device (U.S. Patent No. 974,415). In early September 1910, John and Regina left on a belated honeymoon, a four-month-long cross-country trip by train that included travel to the Sierra Nevada, through the Pacific Northwest, then to Chicago, and on to New York City. During the second week of October, John attended the Belmont International Aviation

Tournament at the Belmont Park racetrack on Long Island, New York. His presence there was noted in the *Washington Post.*[27] Also in October, he and Stanford University president David Starr Jordan were invited to be honorary members of the fledgling Santa Clara Valley Aero Club.[28]

After a lengthy illness, Octave Chanute passed away in Chicago on November 23, 1910. With his generosity, his leadership, and his championing of the free exchange of ideas and the latest technology, he may have done more than any other person to help usher in a new era in human endeavor in aeronautics and aeronautics as a discipline of science. In the later years of his life, he became justifiably recognized as the "father" or "dean" of American aviation. With the death of Langley in 1906, and now Chanute, John felt a sense of responsibility to represent the altruistic attitudes of his generation of enthusiasts from the 1893 meeting in Chicago.

CHAPTER 16

Evergreen

No matter how much one might know of the subject of aviation, in the presence of Professor Montgomery the inclination was all to listen and not talk.

James E. Plew, 1911

In the winter of 1910–11, Edmund Andrews and Raymond Acre spent time together at the Andrews family home at Daytona Beach, Florida. There they conducted additional manned flight tests with *The Lougheed-Acre Replica* via towed flights, free flights off sand dunes, and unmanned flights as a ballasted tethered kite. Andrews marveled at the very flat gliding angle the craft would assume when the tow rope was slackened.[1]

Meanwhile, Montgomery's partners in Chicago were becoming all too familiar with the difficulty of establishing an aviation business. In Illinois and adjacent states, manufacturing companies were expected to sponsor and organize exhibitions; "even flying schools relied upon public performances for a regular income."[2] A succession of companies sprang up in Chicago between 1911 and 1916, but few continued in business for more than a year, and many disappeared without ever building a machine.[3]

Montgomery, Cornelius Reinhardt, and a new assistant, Joseph Vierra, began construction of a new Montgomery monoplane glider, intent on carrying out a new series of gliding experiments in California.[4] Heeding advice from his brother Richard, John wrote to Lougheed and Plew requesting that the patent rights be reassigned to him owing to their lack of performance. Lougheed acknowledged the difficulties of the aviation business and that his own distractions with consulting and writing were a central cause of the failed contract. He returned the patent rights to Montgomery in March.

Early that same month, a new phase of the electric rectifier patent lawsuit got under way in court. Montgomery's attorney established him as an expert in physics and mathematics. A newspaper correspondent following the trial noted, "The defense began its case today by putting Professor Montgomery on the stand and all day he has set forth the secrets of electrical invention and research in a way that has kept counsel grasping for the intricate meaning of the subjects into which the professor delves as though reading from a primer."[5] Despite the draining experience of the protracted legal proceedings, there were moments (albeit rare) of levity that revealed John's wit and sense of humor. Regina C. Montgomery recalled:

> Incidentally much of the aeroplane patent was brought into the case and the testimony abounded in technical phrases. "Parabola," "hyperbola" and other mathematical nomenclature ran through the hours. The opposing council in cross-examining Montgomery sneeringly referred to "the parabolic, hyperbolic, diabolic element introduced into this case." Mr. Montgomery retorted: "I am responsible for introducing the parabolic and the hyperbolic element. I took you to introduce the diabolic element into the case."[6]

The case lingered into May, resulting in a judgment in Montgomery's favor on May 25. However, the lawsuit had depleted nearly all of the funds that John had received through the sale of the electric rectifier. After the court's decision, a wealthy school friend from John's time as a graduate student at St. Ignatius College, Edward McGeary, advanced him $2,500 so that practical experiments with his new glider could continue. This money covered the costs of all materials and salaries for his mechanic (Reinhardt) and pilot (Vierra).[7]

The resulting Montgomery monoplane was based loosely on the third small model glider design from 1903.[8] It incorporated a completely articulated wing, whose angle of incidence could be changed ±5° on the central third, while the panels on the outer third could be deflected downward 7 inches at the trailing edge. The plane also had a fixed fin and stabilizer. The aircraft was controlled solely through use of the rubberized cambric wing.[9] The glider was made for the purpose of exploring airfoil efficiency and its capability for soaring flight, and to minimize the amount of motive power required to sustain powered flight. Initially John incorporated two control sticks, one for each of the pilot's hands. The plane had a wingspan of 26½ feet, a chord of 79 inches, and an empty weight of 145 pounds. The fuselage was constructed of bamboo longerons braced with wiring and steel tubing, the wing spars and ribs were of spruce, and the control mechanism was made of steel, aluminum, ash, and spruce. A four-wheeled undercarriage supported the glider. The aircraft's weight made it impractical for towed launches or self-launching and required another rail-launch facility.

In June 1911, the ASNY appointed Montgomery to the technical board's "research committee." While notifying him of this honor, ASNY member Hugo Gibson said, "It is sincerely hoped that your interests in the development of the aeronautical art, combined with your special facilities for arriving at scientific conclusions securing scientific data in many branches that are at present uninvestigated, will result in your taking hold of the work of the technical committee in an active manner."[10] Also in the summer of 1911, John planned a course of instruction on aeronautics at Santa Clara College.[11] The college was thus poised to be the first educational institution in the western United States to include an aeronautics course as part of its curriculum.[12] In the spring of 1911, John was elected vice president of the Santa Clara Valley Aero Club.[13]

Inspired by the work of Chanute, Lilienthal, and Montgomery, a group of Swiss-based gliding enthusiasts led by David Deluz formed the Club Genevois d'Aviation (Geneva Aviation Club [CGA]) in 1909. In 1910 and 1911, they experimented with several gliders (or "sailplanes," as they termed them) as an inexpensive way to become familiar with flight. Members of the club flew from the slopes of Tivoli and Salève near Bossey, Switzerland.[14] Deluz constructed his own replica of the Montgomery tandem-wing glider design. Dubbed *The Deluz-Montgomery* by the CGA, it was similar in overall appearance to *The Santa Clara,* but given that it was constructed only from photographs, it differed in several ways. For example, its wings were set at a dihedral angle and had flaps similar to those of the French *Antoinette VII* monoplane, although extant photos of Deluz gliding in the aircraft suggest that he also may have used weight shifting for lateral balance.

On September 3, 1911, the GAC organized a "Soaring Competition" at Salève, a small mountain in the French Alps just across the border in France, featuring experimental flights with gliders inspired by Montgomery, Chanute, and Lilienthal designs, as well as other glider types. As the point of takeoff was in France and the landing zone was in Switzerland, these can also be regarded as the first "international" glider flights in history. After the program, Swiss engineer Emile Rieben of the CGA concluded, "My best memory of flight remains the tests on [the] Montgomery, which, in my opinion, was the most powerful sailplane of then."[15] Shortly after the glider contest, the club began selling gliders under the name Sporting Society, Industrial and Commercial (Société Sportive, Industrielle et Commerciale). Based on the initial experimental program, *The Deluz-Montgomery* was chosen as their main glider design, and in 1912 it was renamed *The Geneva,* perhaps to help avoid any possible patent infringement issues. The design became a popular glider. It is unknown whether Montgomery was ever aware of these duplicates in Switzerland.

As a new phase of enthusiasm in aviation was experienced in America and Europe, the Wright Brothers continued to obtain injunctions against other flyers, most notably John Moisant, an American aviator and proprietor of a barnstorming troupe (the Moisant International Aviators) and Thomas Sopwith, a British aviator who ran the Sopwith School of Flying at Brooklands, New York. These legal maneuvers received widespread attention, and Montgomery's patent consequently became an occasional topic of discussion in the press.[16]

Given the legal landscape of the period, and under the assumption that James Plew still maintained control of the Montgomery patent, in mid-October Glenn Curtiss wrote to ask Plew how he might be able to operate under or assume control of it. "I am planning to increase the scope of the aeroplane company," Curtiss explained, "and write to ask what lines arrangements [*sic*] could be made to secure the rights to it, in such manner that infringers could be prosecuted and manufacturers could be licensed."[17] As Plew had abandoned aviation as a business proposition, and Montgomery had resumed control of the patent rights, Plew forwarded Curtiss's letter to John at Santa Clara College. There it waited in a pile of mail while John conducted experiments with the monoplane. Ironically, one month earlier, on September 26, John had placed an order for a Curtiss engine through the shop of well-known local aviator Clarence H. Walker of San Mateo. (He gave Walker a $100 deposit on the $500 motor.)[18] His exact reason for purchasing it remains unknown, but he may well have intended to fit one of his propellers to the Curtiss engine for application to a future model based on the monoplane design.

After surveying the rolling foothills of Mt. Hamilton to the southeast of San Jose, John identified a suitable hillside in the Evergreen District. He requested permission from the ranch owner, John Ramonda, to conduct experiments with kites on a remote section of his land. Ramonda gave his permission but was perplexed as to why an adult man would ever want to experiment with "kites."[19] In mid-October 1911, Montgomery, Regina, Reinhardt, Reinhardt's wife, and Vierra established a camp at the base of a northwest-tending ridge with a U-shaped westerly-facing bowl. At the crest of the ridge, they constructed the launching ramp for the glider. Reinhardt's wife and Regina alternated cooking duties; Regina timed all of the flights. Inspired by their pastoral setting, John christened the glider *The*

Fig. 34. Joseph Vierra ascending in *The Evergreen* from the end of the launch rail at the Ramonda Ranch, Evergreen District, San Jose, Calif., October 1911. (Courtesy Santa Clara University Archives and Special Collections)

Evergreen. After some initial flights, he had Reinhardt replace the vertical dual stick controls with a single steering yoke that John had designed to control both pitch and roll.

An orchard at the base of the hill restricted the maximum forward distance to about 175 to 275 yards.[20] Because of this, after launching and gliding for a short distance, Vierra and Montgomery would bank the glider to the right or left to prolong their glides. Approximately fifty-six flights were made at the facility, with Montgomery and Vierra taking turns as pilot. So efficient was the design that on some flights it was necessary to land prematurely in order to avoid the trees and fences at the far end of the Ramonda tract.[21] Reinhardt later noted, "The experiments at Evergreen demonstrated all and more than Montgomery expected."[22]

As confirmed later by Reinhardt, John made incremental changes to the curvature of the wing through adjustments to splices in the wing ribs. On one flight he was able to gain enough elevation that with a banking turn he landed on top of a second ridge near the point of takeoff. Of this particular flight, Reinhardt noted:

> It was somewhat late but still plenty light and he said "let's try another wing-curvature" which we did. In this instance he now sawed and cut 4" off the wings in front and added 4" onto the back. This flight was a beautiful one. He glided down off the runway into the wind and rose and circled above the tent where his wife, three-quarters of a mile down the hill, was watching him from our supply and repair tent. He then turned

Fig. 35. John Montgomery landing *The Evergreen* in 1911. Note the low angle of approach and that the stick controls had not yet been changed to a yoke control. (Courtesy Santa Clara University Archives and Special Collections)

> around in the air with the wind and rose beautiful as a seagull, directing his flight by means of his makeshift but workable controls, up and back to the ridge from which he had started.[23]

On the morning of October 31, 1911, Vierra completed two flights, and John made another one. At about 9:00 A.M., John prepared for yet another flight. Vierra later revealed that for some reason John appeared nervous. Vierra was aware that John occasionally suffered from vertigo, and he therefore advised him not to make any more flights.[24] Anxious to evaluate an adjustment to the wing ribs, however, John insisted. Just after liftoff, he lost control of *The Evergreen,* and it went into a stall and crashed.

Testimony given by Reinhardt and Vierra eight years after the accident differs somewhat as to what exactly happened. Reinhardt stated that after leaving the track, "I saw him going back with his neck and throwing up his hands." Reinhardt, who was standing about forty to fifty feet away, noted that "the machine was still going ahead and without a control and then she turned to [the] side and he fell down."[25] Vierra, who was standing farther away than Reinhardt, said that Montgomery lost control of the glider just after leaving the launching rail, at an altitude of about fifty feet. After leveling off, "he relaxed the control, reeled to one side, grasped at the framework of the machine, glided for some distance downwardly, striking on the point of one wing lightly, and turning to a vertical position, meaning tail on and rear surfaces in the air, remaining in that position."[26]

Regina was looking down at her watch to time the flight and did not see the crash. When Reinhardt arrived on the scene and pulled John out of the fuselage, John asked, "What happened? What's wrong? How's the machine?" to which Reinhardt replied, "How are you?" "Then his wife came up," Reinhardt said. "His cap had fallen off and as I lifted the machine away from him, leaving it upside down I knelt beside him and saw the small hole just above and back a little from his ear."[27] Reinhardt realized that a stove bolt from the frame had penetrated the base of John's skull about a half an inch.

Vierra immediately took John to his tent, placed him on a cot, and rushed to the Ramonda home for help. While waiting for the doctor to arrive, John talked about the accident with Regina and Reinhardt. Reinhardt relayed his comments to the *San Jose Mercury News*, and they were published in the *Washington Post* later that same day:

> The machine seemed under perfect control, when suddenly one of those little whirlwinds which are common in the foothill regions swept along toward me and struck the machine when I was about 40 feet from the earth. The machine swung around with the whirlwind and then got beyond my control. I tried to turn the course of the glider and thought I had succeeded when suddenly the machine turned half over and dropped 20 or 30 feet from the ground where the final fall came, and it did not seem that the fall would be serious at that height.[28]

On his way to the site, Dr. J. I. Beattie made a wrong turn and was delayed. He arrived at the tent nearly two hours later, at which time he determined that John's injuries were severe enough that he had been beyond medical aid from the time he struck the earth. John J. Montgomery died approximately three hours after the accident. Beattie determined that the cause of death was a hemorrhage of the lungs and the resulting pressure on the heart, as well as hemorrhaging of the brain.[29]

As is typical of such events, theories circulated in the press about the circumstances surrounding John's fatal accident. Richard Montgomery, having visited the Evergreen camp to help with getting the glider to the top of the hill to prepare it for launch, noted the following morning at the coroner's inquest that his brother's exhaustion after repeatedly towing the craft uphill to the launching point had been a contributing factor to the accident. However, newspaper reports based on statements by Vierra and Reinhardt in the days following the crash almost universally attributed John's loss of control of *The Evergreen* to a strong whirlwind. Whether exhaustion and vertigo were contributing factors will never be known.

As news of Montgomery's death spread, public reaction was swift. On the very evening of the accident, the Pacific Aero Club and the Aero Club of California held special sessions and offered statements of condolence to Regina. That same night, Professor Twining and the Aero Club of California issued a proclamation in honor of Professor Montgomery. Eulogies appeared in newspapers and the editorial pages of aviation journals. John J. Montgomery was acknowledged as a well-respected scientist, a pioneer not only of flight but of the scientific discipline of aeronautics.

Thomas Baldwin learned of John's death on November 1. He immediately wrote to Orville Wright to share the news and to reaffirm their mutual attitude toward Montgomery. "Thus ends the career of a man who claimed so much and did so little," he noted.[30] Orville responded, "Montgomery had a number of admirers, but for what reason I never clearly understood, for I cannot think of anything of any value that originated with him."[31] The opinions of Montgomery's two chief adversaries contrasted starkly with those expressed by the overwhelming majority of Montgomery's peers and the general public.

The following week, James Plew told the Aero Club of Illinois: "We have not yet recovered from the shock of Professor Montgomery's death. It came like a thunder clap from a clear sky." The president of Santa Clara College, Father James P. Morrisey, provided an eloquent expression of his valued intellect:

> In him the faculty of Santa Clara loses one of its most honored and devoted members. I feel that by the untimely ending of a remarkably brilliant life the science of aeronautics has sustained a loss that many years will not repair. Professor Montgomery was no mere experimenter with wings. A profound mathematician as well as a keen observer and a persevering thinker, he elaborated his principles and treated them mathematically before experimenting practically.[32]

Public comments, too, were overwhelmingly favorable regarding Montgomery and his life in science. There were plenty of people who found ample reason to admire John J. Montgomery, and they expressed their feelings eloquently.

The funeral was held on November 3, 1911, at Oakland's St. Francis de Sales church. The well-attended service included a requiem mass led by the Reverend Richard H. Bell. Father Morrissey delivered the eulogy. John was interred in the Cleary family plot at the Holy Cross Cemetery in Colma, San Mateo County. The fall of 1911 saw the passing of a true pioneer in the solution of human-controlled flight and the dean of aeronautics in the western United States.

Epilogue

John Joseph Montgomery made outstanding contributions to aviation by his early research into the nature of the laws of flight, by building and testing a series of gliders, by developing improved methods of glider control, and by bringing widespread attention to aviation by the public demonstrations of his glider.

United States National Aviation Hall of Fame, 1964

One unfortunate circumstance arising from John's tragic death was the loss of primary information concerning his research and development in science. During his various relocations over the years (Oakland to San Diego, San Diego to Alton, Alton to Santa Clara), papers and personal effects had been left behind or lost. In the months after his death, a devastated and distracted Regina attempted to move through the grieving process by selling the house and moving into an apartment. Well-meaning friends and relatives sorted through John's personal effects, packed what they thought was relevant to Regina, and donated, threw out, or abandoned what they felt was irrelevant. In an interview years later, Regina described this process as chaotic and acknowledged that a lot of information was lost as a result. The same fate awaited the voluminous and historically interesting correspondence maintained by Zach in the aftermath of his death. Surprisingly few of both Zach's and John's writings were preserved.

Following John's untimely death, however, the Montgomery heirs wanted to ensure that his reputation was preserved and his patent rights were protected. Remembering that Glenn Curtiss had acknowledged the value of the Montgomery patent and had expressed an interest in manufacturing under it in 1911, the heirs moved to challenge his production of aircraft. With Regina Montgomery's brother Alfred J. Cleary acting as attorney for the family, the heirs filed suit against Curtiss in the Southern District of California in 1912, alleging patent infringement.[1] Curtiss's attorneys filed a demurrer on the basis that they were not under the jurisdiction of the California courts. The case was dismissed, and the family lacked sufficient finances to pursue the matter further.[2]

In April 1914, Glenn Martin approached Richard Montgomery at his Oakland office and inquired as to what steps could be taken to protect the Montgomery patent.[3] By that time, Martin had already established himself as a successful aircraft designer in Los Angeles. Over several weeks of meetings, he made a proposal. Alfred Cleary recalled, "[Martin] was anxious to avoid paying royalties on what he regarded as a valueless, or nearly valueless [Wright] patent, and stated that Frank A. Garbutt, his friend and business partner, was willing to take issue with the Wrights provided that he were fortified with an agreement entered into between Garbutt and the Montgomerys respecting part of the proceeds of any sale that might be consummated later."[4]

Garbutt was a wealthy Los Angeles–based industrialist, inventor, and sportsman engaged in auto racing, aviation, and motion pictures. At Martin's urging, Richard and Alfred Cleary met with Martin and Garbutt in Los Angeles for further discussions. Garbutt indicated that he was a longtime admirer of John Montgomery's work and wanted to "see that Prof. Montgomery get what he regarded as his right, namely, recognition as being the inventor of aviation in heavier-than-air machines."[5] He also suggested that with the Montgomery and Lamson patents in hand, he could force the Wrights to make an offer to buy the patent rights and form a combination of the patents of the Wrights, Curtiss, Lamson, and Montgomery.[6]

The heirs felt a sense of urgency about achieving closure on the issue, as Ellen Montgomery was eighty-six years of age and in failing health. Although Cleary had reservations about associating with Garbutt, Garbutt led Richard to believe that he was the man who, in the summer of 1905, had provided moral and financial support to John after Maloney's death, when he needed it the most.[7] In April and May 1914, Richard urged the other heirs to formalize the agreement with Garbutt, which was accomplished in midsummer.[8]

Over the following months, Garbutt was slow to carry out his end of the agreement, but after continuing pressure from the heirs to show some results, he eventually filed suit against two manufacturers in California: Silas Christofferson (the Christofferson Aircraft Manufacturing Company) and an exhibition aviator named B. B. Lewis. However, he neglected to include Regina Montgomery's name (as the principal owner of the patent) on either complaint, and he also failed to get any of the heirs' signatures on them.[9] The family was actually sympathetic to Redwood City, California–based

Christofferson, as he was interested primarily in instruction rather than in the selling of machines, and they were not predisposed to hamper anyone along the lines of scientific development.[10] As for Lewis, upon learning that he was to be served with a lawsuit, he fled the state. Garbutt served Lewis's sister instead.

In February 1915, Cleary noted that Garbutt had "sent Glenn Martin east for the express purpose of entering into an agreement with the Wrights for the purchase of the Montgomery patent."[11] However, history shows that Martin's trip resulted in his entering into a merger with the Wright Company. In August 1915, Garbutt informed Richard Montgomery of this new alliance, noting an abrupt change in Martin's attitude. Martin now "did not believe the Montgomery patent was at all valuable and that it did not conflict with the Wright patent, and that no consideration would be taken by the Wrights or himself of the Montgomery patent."[12] The Montgomerys realized the underlying nature of the Garbutt-Martin partnership and notified Garbutt in writing that they were canceling their agreement. Officially formed in 1916, the Wright-Martin Aircraft Corporation capitalized on the wartime need for the mass production of aircraft under government contract.

When the United States entered World War I in 1917, the owners of early patents for airplanes and related goods charged exorbitant royalties for their use. Aircraft production in the United States nearly came to a halt as airplane producers sued each other for patent infringement. As a result, Congress passed the Naval Appropriation Act of 1918, which included $1 million for the purchase or pooling of airplane patents.

Alfred Cleary and Richard Montgomery envisioned a new strategy. Cleary obtained expert advice in San Francisco from City Attorney Percy V. Long (assistant general counsel for the National Board of Underwriters), City Engineer Michael M. O'Shaughnessy, and Malcolm A. Coles, who had served as assistant U.S. attorney general in the Woodrow Wilson administration and had been in charge of patent litigation against the United States in the Court of Claims. The three experts were all in agreement on the merits of the Montgomery patent and its pioneering nature. Letters in support of this opinion were forwarded to Secretary of the Navy Josephus Daniels and William F. Durand, head of the National Advisory Committee for Aeronautics (NACA). Durand responded to O'Shaughnessy, in part, "I have agreed with you regarding the general character of Montgomery's pioneer work in aviation and I believe he should, in equity, receive some consideration."[13]

On Long's advice, Cleary made a claim through Secretary Daniels against the $1 million congressional appropriation.[14] O'Shaughnessy wrote to Daniels that Montgomery's was "the only basic aeroplane patent, and that all aeroplanes now manufactured infringe on said patent."[15] He recommended that the government purchase the patent. Daniels forwarded these opinions to Charles D. Walcott and J. G. Coffin at NACA, but without any result.

Cleary also presented the claims to the assistant secretary of the Department of the Interior, Alexander T. Vogelsang, as well as to President Woodrow Wilson himself through Wilson's personal secretary, Joseph P. Tumulty. Vogelsang acknowledged the validity of the Montgomery claims.[16] Wolcott, however, remained uncooperative, telling Cleary that he needed to approach the Wright Company and negotiate with them directly. Considering this to be useless advice, Cleary instead sought the support of Senator James D. Phelan (a former classmate of John's at St. Ignatius College) and Vogelsang. They advised that it would be best to have the Montgomery patent adjudicated in the courts. Senator Phelan introduced Cleary to a trusted California-based capitalist who, after consulting with several prominent attorneys regarding the merits of the patent, agreed to finance lawsuits with infringers so long as he would eventually be reimbursed for the court costs as well as 20 percent of any financial awards from the cases.[17] The investor chose to remain anonymous.

On September 22, 1917, the Montgomery heirs filed two lawsuits. The first was against the Wright-Martin Aircraft Corporation (Equity No. 14-298, Southern District of New York) for the manufacture of aircraft for the U.S. government that infringed on the Montgomery patent. The second was against the government itself (Equity No. 33852), as it owned the pooled patents on the infringing aircraft.

In January 1919, depositions were taken in California. Coincidentally, the Wright-Martin Aircraft Corporation of California was dissolved on January 11, 1919, only three days before depositions were to be taken. Although the heirs had notified Garbutt about the cancellation of their agreement three years earlier, he had never formally reassigned his portion of the patent rights to the heirs. When the Wright-Martin attorneys discovered this, they appealed to the court, claiming that the Montgomery heirs did not have full ownership of the patent involved in the lawsuit. Further complicating matters, Garbutt owned stock in Wright-Martin, and therefore had some financial interest in both the plaintiff and the defendant. Judge Hand, who was presiding over the case, sustained the point and suspended judgment, with the consent of the plaintiffs' attorneys, as it was considered impossible to negotiate with Garbutt to get him to relinquish his portion of the patent rights.[18] The Wright-Martin case was dropped in 1921 by mutual consent. Garbutt soon surrendered his rights in the Montgomery patent. The case against the U.S. government continued until 1928.

The complexities of the case go far beyond the scope of this epilogue. However, a central challenge for the plaintiffs was the scarcity of documentation from John's early years, before he went public in March 1905. In addition, the government's legal counsel injected a great deal of bias into the proceedings. In 1921, the government's legal team compiled a document to be used in closing arguments titled "A Summary of Montgomery's Work." As part of their research, the Wright-Martin attorneys had interviewed Orville Wright and Thomas S. Baldwin for their input, and they cited Orville on various points in their dissertation.[19] This document was presented in the court proceedings as a chronology of Montgomery's activities and a critique of some of the plaintiffs' witness testimony. Not surprisingly, the summary was distinctly biased, and rested largely upon misleading statements and errors. They went so far as to characterize Montgomery's efforts in aviation as mere "aeroplane hobbies."[20] In 1919, the government's attorneys also sought out and received an analysis of the Montgomery tandem-wing design from Jerome C. Hunsaker, an aeronautical authority who worked in the navy's aircraft division. Among the court exhibits that the government's attorneys had at their disposal were firsthand accounts describing the actual performance of the Montgomery aeroplane, as well as the testimony of David Wilkie, who had repeatedly flown a Montgomery tandem-wing craft. Additionally, *The Santa Clara* had been reconstructed and was presented in the court case for the depositions taken at Santa Clara University. It is unknown whether Hunsaker was aware of these sources, but he restricted his investigation to the Montgomery patent drawing and, on that basis alone, came to an almost identical conclusion to Orville and Wilbur Wright's contention in 1905 that the Montgomery design was deficient in control, spirally unstable, and dangerous.[21] At the time of both Orville's and Hunsaker's consultation with the government's legal counsel, they were interacting as co-members of the National Advisory Committee for Aeronautics.

During closing arguments, the court was presented with the defense counsel's "Summary of Montgomery's Work." Chief Justice Thomas B. Booth prefaced his final decision in the case in 1928 with his own summary of Montgomery's activities in aeronautics. In the process, he repeated factual errors from the defendants' summary, introduced errors and misinterpretations of his own, and largely ignored the plaintiffs' exhibits and testimony:

> The courts have uniformly held that to show anticipation as against issued letters patent some drawing, some model, some positive means of identification must appear (they indicate the standard that must be met). Oral testimony is regarded as insufficient and unreliable for this purpose. . . . however sincere and factual the testimony—it [oral testimony] will not meet the requirements. . . . At best, the evidence is probative on the single point that the patentee did on the dates stated do the things described, and discloses only the happening of the chronicled events. . . . Montgomery's articles and addresses printed in the record found publication years after his application for the present patent had been filed and granted, and are purely ex post facto (this is a crucial point).[22]

A fundamental flaw in Judge Booth's conclusion was his assertion that the concepts described in the Montgomery aeroplane patent originated in late 1903. Booth showed that many previous patents and publications represented prior art (much of it post-1880s). With this he noted, "In view of the extent of the prior art evidenced by granted patents and the innumerable publications part of the prior art, a subject much too voluminous to discuss in detail, it seems to us idle to contend that Montgomery was a pioneer in this particular field."[23] This statement was subsequently excerpted in certain publications as a broad historical interpretation of Montgomery's importance in aviation.

Judge Booth's comments flew in the face of expert testimony for the plaintiffs. For example, Alexander Klemin, who was an expert in rotary wing design and a leading aeronautical authority in America at the time, gave the following statement during the case in January of 1919:

> It is really extraordinary how much Montgomery knew of the fundamental principles of aerodynamics at this early stage. . . . He possessed substantially a great deal of the knowledge which we do after lengthy and refined investigations. Some of the valuable features already described indicate this, such as the equivalent of a stabilizer, of the modern adjustable stabilizer of fin action, of aileron action, etc. The following sentence in his patent is worthy of note: "A wing is a specially formed surface placed in such a position as to develop a rotary movement in the surrounding air." The presence of this rotary movement is not, as a rule known to even experienced engineers, although such eminent men as Kutta, Jukowski, Prandl, and Lancaster [*sic*] have deduced its presence from theoretical

considerations. . . . In confirmation of Montgomery's statements, refined wind tunnel experimentation by means of pressure measurement over the entire wing surface carried out by both the National Physical Laboratory and Eiffel, indicates that there is a greater pressure at the center of the wing and a greater efficiency there. It is probable for this reason that wings are made to have a large aspect ratio or ratio of span to chord. It is for this reason also that wings are narrowed down at the edges, as Montgomery has done in his patent. And in this and other features, Montgomery shows a precocious knowledge of the art.[24]

Following Judge Booth's decision, the Montgomery heirs were devastated. Exhausted emotionally and financially, they had no choice but to let the patent litigation drop.

Despite this legal interpretation, however, John Montgomery's reputation remained intact. As the decade-long court case played out, authors and historians of aviation history continued to regularly acknowledge his achievements as a pioneer. In 1919, the San Francisco Board of Supervisors adopted a resolution to change the name of the Marina Flying Field (at the time, just east of Crissy Field) to Montgomery Field, "in honor of the late Professor John J. Montgomery, pioneer in the science of aeronautics in California."[25] In 1920, Montgomery Field was established as an airmail field. In 1924, Santa Clara University established the Montgomery Laboratories, the first facility built specifically for engineering research and education. Six years later, an upper story was added for classrooms and drafting rooms. It was built at the site of the present-day Mayer Theatre. In 1925, the National Aeronautic Association designated John Montgomery as Honorary Member No. 6, "In recognition of his distinguished services in aeronautical development."[26] On March 18, 1934, Santa Clara University held a celebration to commemorate the fiftieth anniversary of Montgomery's 1884 glides. Charles Derleth, Jr., a distinguished engineer and dean of the College of Engineering at the University of California at Berkeley, read a paper on Montgomery at the event.

Following World War I, the Treaty of Versailles curtailed Germany's use of powered aircraft. As a result, Germans began experimenting heavily with gliders, pushing the limits of aircraft efficiency. The first international glider meets occurred in 1922 at Clermont-Ferrand, France, and in 1923 at Sussex, England. The sport of soaring arrived in the United States in 1928 through the efforts of three German sailplane pilots, including Peter Hesselbach, who made an exceptional soaring flight of over four hours at Truro, Massachusetts, on the shores of Cape Cod that summer. After this, gliding experienced a resurgence on both coasts. On the West Coast, San Diego became a focal point in this movement owing to the leadership of William Hawley Bowlus and other designers who tested their new sailplanes in southern California.[27]

After the conclusion of the court cases in the late 1920s, the 131 exhibits used by the plaintiffs were returned to Richard Montgomery. Richard passed away in early 1932, before any effort could be made to organize the material. It was then redistributed to Jane and Mary, as well as Regina, all of whom lacked Richard's more comprehensive knowledge of John's research and development in aeronautics.

With the new interest in gliding in the United States, the Montgomery family began receiving requests for information on John's gliding activities. Authors such as Theodore "Ted" Bellak, John B. Crane, Winsor Josselyn, John V. Hettich, and Everett V. Church were among those who came seeking firsthand recollections for the books they were planning to write. The surviving family members were all advanced in age by then, and although they could provide general background information, they found it difficult to remember details that would be useful to engineers. In 1943, Montgomery was also featured prominently in a "History of Aviation" segment in the Walt Disney animated movie *Victory through Air Power.*

As San Diego's economy grew increasingly reliant on aviation throughout the 1930s and into World War II, the community became interested in learning more about the local contribution to aviation history. A movement was soon under way to memorialize John Montgomery, starting in 1944 with an effort to collect and organize original source material, toward the goal of establishing a monument to John's first glides at Otay. Members of the San Diego Junior Chamber of Commerce formed the Montgomery Memorial Committee and initiated correspondence with the Montgomery family and John's former associates. Local glider pilots organized by Jim Spurgeon formed the John Montgomery Glider Club in San Diego, and the Associated Glider Clubs of Southern California established the Montgomery Trophy for the champion of the popular annual Pacific Coast Midwinter Soaring Championships held at the Torrey Pines Gliderport.[28]

Concurrent with these activities, Albert Arnhym, an aeronautical engineer, announced his intention to give a lecture on Montgomery's career to the San Diego chapter of the American Institute of Aeronautics and Astronautics (AIAA), based on his own examination of the primary source material. Reuben H. Fleet, the head of the Consolidated Aircraft Corporation, was also a member of the local AIAA chapter. More importantly, he was a friend of Orville Wright. Knowing Orville's attitude toward the subject of priority in aeronautics, Fleet felt compelled to "handle" this situation personally. Although it came at a time when a larger effort was being made to acknowledge Montgomery, Fleet's reaction to the planned lecture inadvertently triggered a response by Orville Wright and his chief advocates that would have long-term ramifications for Montgomery's legacy in the aviation history literature. On October 3, 1944, Fleet wrote to Orville, expressing concern that such a discussion would not be in Orville's best interest:

> Since the records [on Montgomery] show very early use of the ailerons and other developments, which have proven important, I feel if this matter is not very carefully handled, it might re-open old controversies and wounds. . . . I would greatly appreciate any guidance you might give me so that I can advise the local group soundly, helping them to avoid honoring Professor Montgomery for inventions for which he should not be honored and, above all, to avoid in any way detracting from the honors so rightly bestowed upon my worthy friend, your good self, and Wilbur. . . . Would it be too burdensome for you to dictate a brief summary of the things for which professor Montgomery should be rightly honored . . . could you give me some idea of the arguments for and against him?[29]

Orville's response to Fleet came in the form of a multipage memorandum filled with misinformation, contradictory talking points, and misleading statements apparently intended to present Montgomery's activities and accomplishments in an artificial context and debase his legacy.[30] He had drawn his statements, he said, from a review of "the court record" (*Regina C. Montgomery et al. v. United States*). It would be more accurate to say that he had paraphrased and extrapolated considerably from the government attorneys' "Summary of Montgomery's Work." While the memorandum is far too voluminous to include in its entirety here, some of Orville's main points were as follows:

1. The design of Montgomery's 1903–1905 tandem-wing glider was inspired by Langley's idea for tandem wings, and it also adopted the Wright control system; thus Montgomery's control system was infringing upon the Wrights' patent.
2. Montgomery's aerodynamics theories led him to incorporate "dangerous" features into the design of his machine. These features had been the cause of Maloney's death, and also of Montgomery's death.
3. Montgomery's representation in the aviation literature was based on a fabrication. "What fame attends the name of John J. Montgomery in aviation history is due almost altogether to propaganda put out by

Victor Lougheed. . . . I am convinced that it was he who led the Montgomery Family, apparently a group of sincere and loyal members, into believing that there had been a great and profound thinker in their family, and that his contributions to aviation had been unappreciated and ignored by mankind. . . . Lougheed instigated and directed the lawsuits."

4. The final decision in the court case had confirmed that Montgomery made no contribution to the field of aviation.

5. "The contention of Lougheed and the Montgomery Heirs that Montgomery was the pioneer to whom modern aviation is chiefly indebted was clearly disposed of when the Court said: 'it seems to us idle to contend that Montgomery was a pioneer in this particular field.'"

Significantly, on the final page of Orville's memorandum, he established the context for the soon to be ubiquitous "Montgomery Myth." "Was Montgomery really a pioneer who antedated Lilienthal?" he asked. "Did Montgomery really accomplish a single glide in his early experiments?" In answer he asserted, "The only *reliable* information on Montgomery's early work came through Chanute." (In other words, Chanute had provided the only authentic account of what happened at Otay in the 1880s, and thus Montgomery's own firsthand accounts of these same flights were all fallacious.)[31]

After receiving Orville's memorandum, Fleet met with Arnhym and helped him "correct" his story. Given Fleet's stature in the local aeronautics community and Orville's stature on a broader scale, Arnhym altered the speech he was preparing for the AIAA chapter meeting on November 1, 1944. His delivered lecture characterized Montgomery as an "amateur."[32]

Prior to 1944, the *San Diego Union* had published occasional articles that tended to be positive or neutral toward Montgomery's place in aviation history. Within a few weeks of Fleet's receipt of Orville's memorandum, however, the paper began to feature interviews with citizens who were incensed at the idea that Montgomery should be honored at all. Roughly three decades after his death, John Montgomery had begun to be treated by some as "controversial."

Between 1944 and 1946, Orville provided copies of the memorandum to other Wright advocates, including Earl N. Findley, Fred C. Kelly, and Lester D. Gardner. Without questioning the material, Findley and Kelly simply used it in their own efforts to further marginalize Montgomery. Orville was kept apprised of their progress through correspondence and occasionally provided further input.[33]

In early 1946, Columbia Pictures announced that it was planning to produce a feature-length film on John Montgomery's life with the working title "The Great Highway."[34] The original screenplay, authored by Byron Morgan, was based on independent research using both primary and secondary material. Morgan was a former pilot who, at the age of fourteen, had witnessed Daniel Maloney's flight at Santa Clara on April 29, 1905.[35] Much to Morgan's chagrin, after securing the rights to the screenplay, director William Wellman, Jr., rewrote portions of the script and came up with a new title, *Gallant Journey.* Fred Kelly notified Orville about the motion picture in February 1946 and promised to launch a campaign to counter its release.[36] Throughout the production of the film in 1946, Kelly conducted an extensive letter-writing campaign to the management of Columbia Pictures, including Paul Johnson, Eric Johnson, director William Wellman, Jr., Harry Cohn, and screenwriter Byron Morgan, stating repeatedly: "There is no truth whatever in reports that Montgomery contributed in any way to successful aviation, or that he made successful glides back in the 80's."[37] He characterized the project as "an unscrupulous falsification of history."[38] He kept Orville apprised, and Orville, in turn, provided him with additional talking points on Montgomery. In April 1946, Earl Findley, the editor of the aeronautical journal *U.S. Air Services,* authored and published an article titled "The Montgomery Myth."[39] It was copied to the New York and Los Angeles United Press Associates Bureau chiefs by Kelly, and was forwarded by Orville to Wright advocates Lester Gardner and Griffith Brewer.[40] Findley's article was essentially an excerpt of Orville's 1944 memorandum, stated as a simple matter of fact, and without citing Wright as a source. According to the article, mythmaking was profuse throughout Montgomery's career.

Despite Findley's and Kelly's efforts, the movie was completed in September 1946 and debuted in San Diego. John's surviving siblings James and Jane, James's daughters Marie and Elinor, and granddaughter Marcia J. Wendt (co-author Harwood's mother) attended the showing and took part in the groundbreaking ceremony at the site of a planned Montgomery Monument in Otay. They also visited the site of the former Fruitland Ranch. Although James was delighted by the concept of a monument at the location of one of John's first flights, he felt that the broader context of John's "thought and development of the aeroplane" at Fruitland was a more appropriate aspect of the history to memorialize.

The location of the monument itself is interesting, as it was based on James Montgomery's descriptions in correspondence and interviews with Winsor Josselyn in 1939. Josselyn visited the Otay region to determine the location of the Fruitland Ranch and the site of John's first glides. The Montgomery family had always said that the first flights of John's first glider took place at a location known to Otayans as "Wheeler Hill" (part of a large tract owned by P. H. Wheeler).[41] An article from the *San Diego Union* in 1883 described the Wheeler property as "a fine ranch on Monument Avenue, about two thirds of which lies on the high [Otay] mesa, south of the [Otay] Valley, and the balance is in the bottom."[42] Josselyn identified a location that matched James's description of the location of the first glides, and he wrote to James to describe this spot on the northwest edge of Otay Mesa:

> The westerly end [of the mesa], pointing toward the lower end of San Diego Bay, terminates in three points, like a trident. Best suited for gliding down a gentle, long slope would be the middle one which, in 1940, pointed toward a hog ranch on the hillside beyond. Standing on the brow of the middle point, the spectator is looking down about a 10 or 15 degree slope and he finds that the prevailing wind comes up it in fair strength, and I would have chosen that spot from which to make such a[n] experiment as you two did.[43]

In 2006, we conducted our own research to determine the most probable location of these same flights and identified a spur ridge located at the northwest corner of the mesa. In 2008, author Harwood discovered the letter quoted above in the archives at Santa Clara University presenting Josselyn's independent conclusion describing an adjacent ridge located just a few hundred feet to the north.[44]

Sometime between this discovery by Josselyn in 1940 and the purchase of land for the planned Montgomery Monument site (1944–46), the proposed location shifted one mile west of the mesa's edge to the rolling hills below (the site of present-day Montgomery-Waller Park). It is unclear why the top of the mesa was not chosen as the location for the memorial, although the comparative real values of the two properties and/or the availability of the land may have been a factor. Both locations were once part of P. H. Wheeler's holdings.

In 1946, the Chula Vista School District christened their newest school the John J. Montgomery Elementary School. That same year, Santa Clara University installed and dedicated a granite obelisk at the site of Maloney's April 29, 1905, high-altitude flight in *The Santa Clara*. The following year, the remaining portions of *The Santa Clara* and *The Evergreen*, along with a copy of Montgomery's 1894 manuscript "Soaring Flight," were collected by Paul E. Garber, historian emeritus for the Smithsonian's National Air

Museum (later the National Air and Space Museum), for preservation at the Smithsonian.[45]

In late 1949, preparations escalated toward the dedication of the Montgomery Memorial (California Historical Landmark #711) and the renaming of Gibbs Flying Field to Montgomery Field in San Diego. Erected in 1950, the memorial itself includes a silver static test wing panel from a Consolidated B-32 Dominator, which was mounted upright so that it would be visible for miles. (It is a rare surviving portion of a B-32.) The monument was financed by contributions from the San Diego Junior Chamber of Commerce and Columbia Pictures. During the promotion of these activities, articles on Montgomery became more common in the local press throughout the winter and spring of 1950. In response, Fred Kelly launched another round of anti-Montgomery propaganda, with articles written for MIT's *Technology Review* and *Science Digest*. They were yet another unattributed reiteration of Orville's 1944 memorandum.

In 1950, a portion of the new Interstate 5 freeway constructed between San Diego and Tijuana, Mexico, directly bisected the former Fruitland Ranch and was named the Montgomery Freeway. The buildings from the ranch had been gone for more than thirty years. In January 1916, San Diego had experienced torrential rains for nearly two weeks, culminating in a powerful storm on January 25 that caused the Lower Otay Dam to fail. The resulting flood inundated Otay Valley with thirteen billion gallons of water, carrying away homes and killing fourteen people before reaching San Diego Bay some seven miles away. The Fruitland Ranch, including the residence, the outbuildings, and John's workshop/laboratory from the 1880s, were all washed away in the rapids, which scoured up to thirteen feet of topsoil from Otay Valley.

In May 1950, two major events were held in San Diego to honor Montgomery's contributions to flight. On May 20, Gibbs Field was officially dedicated as Montgomery Field with a plaque that stands near the administration building. World War II aviation hero Jimmy Doolittle addressed the crowd. On the following day, the Montgomery Memorial was dedicated at Otay. Paul E. Garber provided the opening address.

Concurrent with these commemorations, a series of articles by Fred Kelly titled "Miracle at Kittyhawk" were published in the May, June, and July 1950 issues of *Atlantic Monthly*; they were expanded into a book in 1951.[46] It includes an excerpt of Orville Wright's 1944 memorandum, with Orville finally identified as the source. Over the years, authors have continued to breathe new life into Orville's premises, including Marvin McFarland (*The*

Fig. 36. Flood damage at the mouth of Otay Valley, after the collapse of the Lower Otay Dam, January 1916. The devastation of Fruitland Ranch is apparent toward the left side of the image (only a few trees and remnants of structures remained). The Swiss Park and Club in Chula Vista currently occupies the land where the Montgomery residence once stood. (Courtesy San Diego History Center)

Papers of Wilbur and Orville Wright, 1953), C. Gibbs-Smith (*A History of Flying*, 1953; *Aviation*, 1960), and Fred Howard (*Wilbur and Orville*, 1987).[47] In each publication, the authors claimed that the "court record" (*Regina C. Montgomery et al. v. United States*) had served as a principal source for their discussions. Gibbs-Smith continued to publish books on aviation history, each time expressing greater skepticism about and dismissal of Montgomery's contribution. The books authored by Kelly, Mc-Farland, and Gibbs-Smith received wide circulation and are still considered by many to be landmark compendiums on the Wright Brothers in the larger context of American aviation history.

In 1961, sailplane designer Stanley Hall and a group of aeronautical engineers from the Lockheed Aircraft Corporation constructed a replica of *The Evergeen* using between 3 and 5 percent of the original components.[48] Team leader and aerodynamicist Richard B. Campi performed a scientific evaluation of the aircraft. With supplemental data on material properties published in 1911 and 1913, the team calculated the required control forces and the movement of the center of pressure relative to changes in the angle of attack, evaluated the wing and the lifting horizontal stabilizer, and estimated the flight speed. The yoke steering wheel was connected to the control system, which operated much like a modern aircraft in concept. Campi noted, "The control system was an ingenious combination of members operated by a control column and hand wheel. By an inventive arrangement of links, pulleys and hinges, all forces applied to the craft in a system in a sound structural manner."[49] From their analysis, the team reached the following conclusion:

> Combining the lift characteristics of the wing and the stabilizer it was determined that the c.g. [center of gravity] location relative to the combined center of pressure always resulted in a very stable vehicle. . . . these considerations may seem routine in the light of today's practice but applied at a time when man took his first steps into the unknown ocean of the air, it indicates that Montgomery was not only a remarkable inventor and engineer, but that he also had a very sound grasp of the man-machine relationship which today comes under the discipline of Human Engineering.[50]

In 1961, Santa Clara County established a monument to commemorate the location of Montgomery's flights with *The Evergreen* at the corner of Yerba Buena Road and San Felipe Road in San Jose. The nearby Montgomery Hill was commemorated with a plaque. It was later designated a California Historical Landmark (#813) in 1967. It is still possible to stand at the top of the hill and look down into the topographic bowl used for gliding flights by Montgomery and Vierra, unchanged since 1911.

In 1962, the National Society of Aerospace Professionals (NSAP) created a series of awards to be given to deserving individuals for their pioneering work in aeronautics and aerospace. On the theme of American pioneers of aviation, the NSAP established the Octave Chanute, Lawrence Sperry, and John J. Montgomery Awards. The Montgomery Award was given in 1962 to Paul F. Bikle, the director of NASA's Dryden Flight Research Center from 1959 to 1971, along with seven X-15 pilots: Major Robert M. White, Neil A. Armstrong, Scott Crossfield, John F. Yardley, Joseph A. Walker, Commander Forrest S. Petersen, and Major Robert A. Rush.[51] Subsequent recipients included Leroy G. Cooper, Virgil I. "Gus" Grissom, Kenneth S. Kleinknecht, Ernest R. Letsch, John B. McKay, Dr. George F. Mechlin, Jr., and Art Lowell, as well as several others involved in NASA's Polaris program. All were leaders in their respective fields. The sponsor of the award, the Institute of the Aeronautical Sciences (IAS) survives today as the AIAA, and several AIAA members have published articles recognizing Montgomery's leadership in American aviation.

John J. Montgomery was inducted into the National Aviation Hall of Fame at Dayton, Ohio, in 1964. His biography there includes this statement: "John Joseph Montgomery made outstanding contributions to aviation with his early research into the nature of the laws of flight, and by building and testing a series of gliders. Montgomery also developed improved methods of glider control, and brought widespread attention to aviation through the public demonstrations of his glider."[52]

In 1967, Arthur D. Spearman, S.J., archivist for Santa Clara University, completed the first in-depth biography of Montgomery, *John J. Montgomery: Father of Basic Flying*. Spearman visited San Diego to promote the book and the local press. At the same time, local San Diegan Waldo Waterman initiated correspondence with Herbert Lockwood, editor of the *San Diego Independent*, challenging all of the facts surrounding Montgomery. Waterman had flown a Chanute-type biplane glider as a boy in 1909 from a ravine near San Diego, and he felt that he rather than Montgomery should be acknowledged as "first to fly" in San Diego. Lockwood then published an article in the *Union* on May 12, 1968, challenging the validity of Montgomery's early work, and making heavy use of Orville's memorandum. Waterman followed up with an interview in the *Los Angeles Times* on June 22, 1969, under the headline "1883 Legend Called 'Hoax.'"

In 1973, Waldo Waterman, Reuben Fleet, and a mutual friend, Norman B. Neely, sponsored an open competition to "recreate the reported Montgomery hang glider flight" of 1883.[53] The official guidelines stated that "many people have felt Montgomery's claim of a controlled flight in the 1880's to be open to question due to lack of suitable documentation."[54] The entrants were restricted to the specifications provided by the contest organizers for what they believed the Montgomery glider of 1884 looked like, the launching methods implied in *Progress in Flying Machines*, and the general construction as depicted by Lougheed in *Vehicles of the Air*. The rules followed Orville's 1944 memorandum in referring to Chanute's interpretations of Montgomery's gliders published in *Progress in Flying Machines* as the only authoritative account on the gliders and their design. They failed to mention primary accounts and the statements made by John Montgomery after the 1894 publication of Chanute's book.

The competing gliders were evaluated on August 25 and 31, 1975. Although the sport of modern hang gliding was popular at the time in California, the response to the competition was lukewarm; only seventeen people paid the entrance fee, and few if any succeeded in demonstrating the power of the machine under the constraints of the competition.[55] However, these "Montgomery Meets" followed the Otto Lilienthal Universal Hang Glider Championships of May 23, 1971 (the first organized event in modern U.S. hang gliding history), and helped lead to the formation of the current United States Hang Gliding and Paragliding Association. Jack Lambie promoted the first Montgomery Meet at the location of the Montgomery Memorial in Otay in August 1971, and hang glider enthusiasts of the early 1970s had an opportunity to experience what it was like to launch and fly using only one's feet for propulsion.

Over time, additional honors were awarded in Montgomery's name. In 1975, the John J. Montgomery Elementary School was completed and dedicated in San Jose's Evergreen district. The U.S. Air Force Auxiliary Civil Air Patrol Squadron 36, based in San Jose, was named the John J. Montgomery Memorial Cadet Squadron. Ken Palmer of Squadron 36 constructed a replica of Montgomery's first glider in 2007–2008. The Experimental Aircraft Association (EAA) established a John J. Montgomery Chapter at San Jose. His 1911 home was designated as a City of Santa Clara Historical Landmark in 1993. In 1996, Montgomery's first glider of 1883–84 was recognized as an International Historic Mechanical Engineering Landmark by the American Society of Mechanical Engineers (ASME). According to the designation, Montgomery's glider was

> the first heavier than air craft to achieve controlled, piloted flight. The glider's design based on the pioneering aerodynamic theories and experimental procedures of John Joseph Montgomery (1858–1911), who designed, built, and flew it. This glider was way ahead of its time, incorporating a single parabolic, cambered wing, with stabilizing and control surfaces at the rear of the fuselage.[56]

In 1975, Congressman Norman Mineta, the former mayor of San Jose (and later the secretary of transportation under President George W. Bush), read into the *Congressional Record* an address commemorating the seventieth anniversary of Maloney's April 29, 1905, flight. In an attempt to reconcile Montgomery's relative obscurity with the anti-Montgomery campaigns, Mineta stated:

There are perhaps many reasons for this unfortunate obscurity, but among them is the bitterness created in a long and hard-fought trial during the 1920's which pitted the Montgomery heirs against some of the proprietary claims of the Wright brothers. The ill will born of that long struggle has been passed on to successive generations of advocates, like some backwood family feud that over many generations has become an unthinking and automatic animosity. . . . I appeal to all scholars to set aside their assumptions on this subject and to each insist on reviewing the facts first-hand, not through intermediate generations of hatred. The debate over proprietary ownership of certain inventions may necessarily admit only of absolute determinations, but in the history of ideas we see more often than not that progress is wrought by the efforts and contributions of many people, and that in history there is room for them all.[57]

In 2002, Montgomery was inducted into the U.S. Soaring Hall of Fame, operated by the Soaring Society of America. According to the society's defining statement, it "recognizes individuals who have made the highest achievements in, or contributions to, the sport of soaring in the United States of America."[58] American gliding pioneers Octave Chanute and Orville and Wilbur Wright are also honored in this hall of fame. Despite all the accolades, the "Montgomery Myth" continues to resurface. In 2005, Herbert Lockwood published a book titled *The Montgomery Myth: The Flight That Never Was,* in which he cited Orville's 1944 memorandum for several of his talking points.

On March 19, 2005, a Centennial Celebration of Soaring Flight, complete with a replica of *The Santa Clara,* was held on the former site of the Rancho San Andreas, near the present-day location of Dolphin Drive in Aptos. A monument was installed by the Ancient and Honorable Order of E Clampus Vitus adjacent to the actual location of the balloon-assisted launches of the glider there. Later in 2006, another monument honoring Montgomery's work at Mount St. Joseph's College was established at the Rohnerville Airport by the AV8CANDO organization. On March 15, 2008, a sculpture in tribute to Montgomery was unveiled in the Evergreen district of San Jose. The 30-foot-tall steel structure representing *The Evergreen* monoplane wing was placed on a 32-foot-diameter plaza designed by San Franciscan Kent Roberts. The following month, on April 5, 2008, the Hiller Aviation Museum in San Carlos, California, celebrated the 125th anniversary of John Montgomery's first glider flights. Most recently, in 2011, the School of Engineering at Santa Clara University, in conjunction with the American Society of Mechanical Engineers (ASME), established a $40,000 endowment fund for the John Joseph Montgomery Gold Medal, to be awarded annually to individuals and organizations for "Distinguished Innovation in Aeronautics."

John Montgomery's life was not lived in pursuit of glory and monetary reward, but in applying himself for the love of science. His controlled investigations and resulting theories contained some basic aeronautical truths. After reviewing Montgomery's 1894 manuscript "Soaring Flight," modern aerodynamicist John H. McMasters described Montgomery as having

> somehow got at two of the toughest, most counter-intuitive factors in correctly explaining how wings and airfoils work: That the flow over the upper surface of a curved plate can be made to follow a convex curve, with a consequent reduction in pressure (relative to the atmospheric pressure surrounding it); and the circulatory motion surrounding the wing section "induced" by its shape. . . . given the state of knowledge at the time on this hugely complex, and largely "invisible to the naked eye" topic, he did an amazing job. . . . He clearly was well educated by the standards of the day and had a good understanding of Newtonian physics in addition to being an excellent observational experimentalist. He got a lot right, and has fairly earned his place in aviation history.[59]

However, perhaps the most potent characterization of John Montgomery was offered by Harry La Verne Twining in a resolution by the Aero Club of California at the time of John's death on October 31, 1911:

> He was above the temptation of power, and rather than to battle in the courts for the mastery of things, he chose to strive for the perfection of his ideals. He had no predecessor for he was the father of them all; his work was original and that of a genius. The world's tempting offers held for him no personal fame for his character had been molded into that of a man who served God rather than a greed for gold. There at the college in Santa Clara he labored with his students not alone as professor, but in that true Christian brotherhood of fellow worker, conscientious and sincere. A scientist in the fullest sense of the word. His life was faithfully devoted to that cause which claims the lives of many and will live forever—SCIENCE.[60]

Those who knew John Montgomery during his life valued his contributions to science and his willingness to share them selflessly with society.

NOTES

Chapter 1. The Allure of California

1. Biographies of Zach Montgomery are available in Joseph E. Baker, *Past and Present of Alameda County, California*, 2 vols. (Chicago: S. J. Clarke, 1914), 2:547–49; Peter J. Delay, *History of Yuba and Sutter Counties, California* (Los Angeles: Historic Record Co., 1924); J. M. Guinn, *History of the State of California and Biographical Record of Oakland and Environs, Also Containing Biographies of Well-Known Citizens of the Past and Present* (Los Angeles: Historic Record Co., 1907), 2:733; *Herringshaw's Encyclopedia of American Biography of the Nineteenth Century* (Chicago: American Publishers' Association, 1902), 668; Charles Lanman and Joseph M. Morrison, *Biographical Annals of the Civil Government of the United States: From Original and Official Sources*, 2nd ed. (New York: J. M. Morrison, 1887), 349; Frank Clinton Merritt, *History of Alameda County* (Chicago: S. J. Clarke, 1928), 198; Alonzo Phelps et al., *Contemporary Biography of California's Representative Men* (San Francisco: A. L. Bancroft, 1882), 181; John Joseph Shanahan, "Zachariah Montgomery: Agitator for State and Individual Rights" (master's thesis, University of California, Berkeley, 1955); Oscar Tully Shuck, *History of the Bench and Bar in California* (Los Angeles: Commercial Printing House, 1901), 537–38; Thompson & West, *Official and Historical Atlas of Alameda County, California* (Oakland, Calif.: Thompson & West, 1878); Wm. H. Chamberlain and Harry L. Wells, *History of Sutter County, California: With Illustrations Descriptive of Its Scenery, Residences, Public Buildings, Fine Blocks and Manufactories* (Oakland, Calif.: Thompson & West, 1879).

2. Land Patent Records, January 1, 1828, St. Louis Land Office. James Evoy purchased 160 acres on Hancock Prairie, Callaway County, Missouri (Township 47N/Range 7W/SW quarter of Section 36), about eighty miles west of St. Louis.

3. John D. Unruh, Jr., *The Plains Across: The Overland Emigrants and the Trans-Mississippi West, 1840–60* (Chicago: University of Illinois Press, 1993), 510 n. 5.

4. Jane E. Montgomery, "A Few Notes and Anecdotes Told by John Montgomery Recalled by Jane E. Montgomery" (unpublished ms., March 1946, 4 pp.), John J. Montgomery Collection, Santa Clara University Archives and Special Collections, Santa Clara, Calif. [hereafter SCU-JJM]. This oral account, based on family tradition as told to Jane by her mother, Ellen B. Montgomery, describes incidents that occurred during the Evoys' overland journey to California in 1849. Supplemented with information provided by modern historians concerning routes, estimated travel times, and landmarks, it provides a rough reconstruction of their trip. The Evoy/McCourtney emigrant party is also referred to in *Message from the President of the United States: In Answer to a Resolution of the Senate, Calling for Further Information in Relation to the Formation of a State Government in California; and Also, in Relation to the Condition of Civil Affairs in Oregon*, May 22, 1850, Sen. Ex. Doc. 52, 31st Cong., 1st sess. (Washington, D.C.: s.n., 1850), 103, 109; Hugh Brown Heiskell, *A Forty-niner from Tennessee: The Diary of Hugh Brown Heiskell*, edited by Edward M. Steel (Knoxville: University of Tennessee Press, 1998), 28, 49.

5. Phelps et al., *Contemporary Biography*, 68.

6. Roger H. Futrell, "Zachariah Riney: Lincoln's First Schoolmaster," *Lincoln Herald* 74 (Fall 1972): 136–42.

7. Zach Montgomery, "A Few Notes on the Family Name and History of Zach Montgomery, Written Down at the Earnest Request of a Beloved Daughter" (unpublished ms., 1884, 8 pp.) [SCU-JJM].

8. John Bidwell, *Journal of a Trip to California in 1841* (Weston, Mo.: n.p., 1844). Bidwell later became a leading citizen and politician in California.

9. Zach Montgomery, "First Thoughts of California," *Family's Defender Magazine and Educational Review* 3, no. 3 (1883): 119–28, Catholic Newspapers in Microform, University of Notre Dame Archives, South Bend, Ind. [hereafter UND].

10. Jo Ann Levy, *They Saw the Elephant: Women in the California Gold Rush* (Norman: University of Oklahoma Press, 1992).

11. Zach Montgomery, "Address Delivered before the Erodelphian Society, St. Joseph's College, on the Anniversary of the College, April 19, 1849," Fordham University, Department of Archives and Special Collections, New York, N.Y.

12. Zach Montgomery, "Remembrances," *Tidings* 6, no. 40 (October 6, 1900), 1–4. Published posthumously.

13. Zach Montgomery, "Personal Remembrances: Sacramento City in 1850," *Family's Defender Magazine and Educational Review* 4, no. 2 (1884): 119–28 [UND].

13. In 1850, Ringgold was a small mining camp and trading post located in a narrow valley alongside Weber Creek, two miles southwest of Placerville, California.

14. Zach Montgomery, "Personal Remembrances: Sacramento City in 1850," emphasis in the original.

15. Ibid.

16. Ibid.

17. Ibid.

18. José Vicente Peralta (1812–1871) was a son of Antonio Maria Peralta (1801–1879). Antonio Maria Peralta built the first non–Native American dwelling in Oakland in 1821, and the Peralta family were major landowners in what is now Oakland. The Peralta family history is preserved at Peralta Hacienda Historical Park.

19. Alameda County Hall of Records, Deeds Book "A," 669. On December 9, 1853, Mrs. B. M. Evoy purchased about ninety-seven acres from Jacob Tewksberry ("Tract No. 36") in the town of Contra Costa, Alameda County (formerly Contra Costa County), about two miles north of Oakland. The price was $3,000 in gold. Included with the property were "agricultural implements including two ploughs, one harrow, hoes and shovel, scythe, rakes and pitching forks, baskets with also a lot of loose Redwood pickets."

20. A tributary of the Pit River in Shasta County was later named Montgomery Creek in Zach's honor.

21. Zach's appointment is documented in the Records of the Clerk Recorder, Sutter County Office of the District Attorney, Yuba City, Calif.

22. Zach Montgomery, "Remembrances."

23. Interment Register, St. Joseph's Catholic Cemetery, Marysville, Yuba County. The register indicates that Helen was born December 20, 1835, in Rochester, New York. See also the obituary notice for Helen Francis Montgomery in the *Marysville Herald*, July 20, 1856.

24. Phelps et al., *Contemporary Biography*; Interment Register, St. Joseph's Catholic Cemetery, Marysville.

25. Zach and Ellen's children were, in chronological order, John Joseph Montgomery (1858–1911), Zachariah Montgomery, Jr. (1858–1861), Mary Clotilda Montgomery (1859–1949), Margaret "Maggie" Helena Montgomery (1861–1931), Eleanor Rose Montgomery (1861–1864), Richard "Dick" Joseph Montgomery (1863–1932), James "Jim" Patrick Montgomery (1865–1956), and Jane "Jennie" Eleanor Montgomery (1869–1955).

26. Mark J. Hurley, *Church-State Relationships in Education in California* (Washington, D.C.: Catholic University of America, 1948), 29–36.

27. Carl I. Wheat, "'California's Bantam Cock': The Journals of Charles E. DeLong, 1854–1863," *California Historical Society Quarterly* 9 (1930): 245–381; 10 (1931): 165–282; 20 (1941): 20; 46 (1967): 154.

28. Harold M. Hyman, "New Light on *Cohen* v. *Wright*: California's First Loyalty Oath Case," *Pacific Historical Review* 28, no. 2 (1959): 131–40.

29. Helen Dare, "Prof. J. J. Montgomery–The Personality of the Man," *San Francisco Call,* May 7, 1905. Based on an interview with John J. Montgomery.

Chapter 2. The Earth and Vaulted Sky

1. Zach Montgomery, *The Poison Fountain; or, Anti-Parental Education: Essays and Discussions on the School Question from a Parental and Non-Sectarian Standpoint* (San Francisco: Z. Montgomery, 1878).

2. Mark Twain, "Democratic Meeting at Hayes Park," *San Francisco Morning Call,* August 3, 1864.

3. Jane E. Montgomery, "A Few Notes and Anecdotes Told by John Montgomery Recalled by Jane E. Montgomery" (unpublished ms., March 1946, 4 pp.) [SCU-JJM].

4. Ibid.

5. Dennis J. Kavanagh, "The Story of the Aeroplane," in *The Aeroplane* (Santa Clara: Aeroplane Advertising Co., 1905), 16–23. Kavanagh was an instructor and a colleague of Montgomery's at Santa Clara College. The article was based on personal interviews with John.

6. Editorial note, *Daily Alta California,* January 3, 1867. See also "The Avitor," *Daily Alta California,* June 16, 1867.

7. "The Avitor," *Daily Alta California,* June 16, 1867.

8. *The Avitor Hermes Jr.* was a test version of what was intended to be a larger, passenger-carrying, steam-driven airship.

9. James P. Montgomery, notes from unpublished interview with Winsor Josselyn, Oakland, Calif., April 8, 1939 [SCU-JJM]. Peter Wude's presence in the Montgomery home in Oakland, along with basic information on his background, is documented in the 1870 U.S. Census.

10. Gerald McKevitt, S.J., *The University of Santa Clara: A History, 1851–1977* (Stanford, Calif.: Stanford University Press, 1979).

11. *Santa Clara College Yearbook, 1874–1875* (Santa Clara: Santa Clara College Press, 1875).

12. Dare, "Prof. J. J. Montgomery."

13. Now the University of San Francisco.

14. John Bernard McGloin, *Jesuits by the Golden Gate: The Society of Jesus in San Francisco, 1849–1969* (San Francisco: University of San Francisco, 1972).

15. Friar Michael Kotlanger (University of San Francisco archivist), email message to Craig Harwood, April 8, 2005.

16. "St. Ignatius Commencement," *San Francisco Bulletin,* June 2, 1877.

17. California College was later renamed the University of California at Berkeley.

18. *Saint Ignatius College Catalog for Academic Year 1878–1879* (San Francisco: Saint Ignatius College, 1879).

19. Ibid.

20. Jane E. Montgomery, "A Few Notes and Anecdotes."

21. Ibid.

22. Ibid.

23. "A Young Genius," *Our Paper* (Berkeley), June 28, 1879.

24. "In those days" refers to the period in the mid- to late 1870s when both Montgomery and Morrison attended St. Ignatius College.

25. Arthur D. Spearman, S.J., *John J. Montgomery, 1858–1911: Father of Basic Flying* (Santa Clara: Santa Clara University, 1967).

26. Joseph W. Riordan, *The First Half Century of St. Ignatius Church and College* (San Francisco: H. S. Crocker Co., 1905).

27. For background on Father Bayma, see "Bayma, Joseph," in Charles G. Herbermann et al., *The Catholic Encyclopedia: An International Work of Reference on the Constitution, Doctrine, Discipline, and History of the Catholic Church* (New York: Catholic Encyclopedia Co., 1907), 2:360. For Father Neri and Father Bayma, see Paul Totah, "The Founding of the St. Ignatius College (1849–1861)," *Spiritus Magis: 150 Years of St. Ignatius College Preparatory, Genesis IV, History Supplement* 41 (2005): 28–31.

28. "The school question" was the proposition to tax citizens in order to maintain a compulsory public education system in California.

29. John remained in Oakland, where he managed the Montgomery Bros. store.

30. Lynne N. Christenson and Ellen L. Sweet, *Ranchos of San Diego County* (Charleston, S.C.: Arcadia Publishing, 2008).

31. Ibid.

Chapter 3. Tutors in the Art of Flying

1. Richard J. Montgomery, in answer to question 4, in direct testimony at San Francisco, *Regina C. Montgomery et al. v. United States*–Equity No. 33852 [hereafter Court (Equity No. 33852)], January 14, 1919.

2. Jane E. Montgomery, "A Few Notes and Anecdotes."

3. James P. Montgomery to Winsor Josselyn, November 18, 1946 [SCU-JJM].

4. Jane E. Montgomery, affidavit, February 3, 1945 [SCU-JJM]. This and other affidavits of the period (1944–46) by Montgomery's surviving siblings were collected by the Montgomery Memorial Committee, administered by the San Diego Jaycees in their efforts to collect primary source statements and honor Montgomery.

5. Ibid.

6. Ibid.

7. Dare, "Prof. J. J. Montgomery."

8. A phenomenon referred to as "ridge lift."

9. John J. Montgomery, "Our Tutors in the Art of Flying," *Aeronautics* (New York) 8, no. 9 (1915): 99–100. This article was originally provided to Ernest L. Jones by Montgomery in 1907, and was published posthumously.

10. James P. Montgomery, in answer to question 7, in Court (Equity No. 33852), January 19, 1919.

11. Victor Lougheed, *Vehicles of the Air: A Popular Exposition of Modern Aeronautics with Working Drawings* (Chicago: Reilly and Britton, 1909), 190.

12. Octave Chanute, *Progress in Flying Machines* (New York: American Engineer and Railroad Journal, 1894).

13. Ibid.

14. Ibid.

15. John J. Montgomery, "The Origin of Warping: Professor Montgomery's Experiments," *Aeronautics* (London) 3, no. 5 (1910): 63–64. Abstract of a lecture given to the New York Aeronautic Society on April 21, 1910. This lecture ("The Origin of Warping") was reprised from a speech delivered by Montgomery at the Conference on Aerial Navigation in Chicago, 1893. The British *Aeronautics* Journal article was later reprinted in "Death of Professor John J. Montgomery," *Aeronautics* (New York) 9, no. 4 (1911): 151–54.

16. Ibid.

17. Louis P. Mouillard, translation in Chanute, *Progress in Flying Machines*, 150.

18. Otto Lilienthal, in *Zeitschrift für Luftschiffahrt und Physik der Atmosphäre*, translation in ibid., 277.

19. "Improved Dynamo Machine," *Scientific American Supplement* 15 (1883): 6247–48. This incident was also related by Jane E. Montgomery in a letter to Winsor Josselyn, March 1946 [SCU-JJM].

20. Richard J. Montgomery, in answer to question 4, in Court (Equity No. 33852), January 14, 1919.

21. Harold E. Salley, *History of California Post Offices, 1849–1990*, 2nd ed., edited by Edward L. Patera (Lake Grove, Ore.: The Depot, 1991).

22. "The Ethics of Invention," *Family's Defender Magazine and Educational Review* 2, no. 1 (1882): 59–65.

23. "Patents for Processes," *Family's Defender Magazine and Educational Review* 1, no. 1 (1881): 505–15.

24. Montgomery filed a U.S. patent application for this method on May 27, 1884, and was granted U.S. Patent 308,189 on November 18, 1884.

25. The known migration patterns of pelicans in the San Diego area and John's statements above date these observations to spring or fall 1883.

26. Montgomery, "Our Tutors in the Art of Flying."

27. Dennis J. Kavanagh, "The Story of The Aeroplane," in *The Aeroplane* (Santa Clara: Aeroplane Advertising Co., 1905), 16–23.

Chapter 4. Wings over Otay

1. Richard J. Montgomery, in answer to question 34, in Court (Equity No. 33852), January 14, 1919.

2. "Successful Flying Machine," *San Francisco Call,* May 1, 1905. Montgomery is quoted extensively in the article.

3. "Machine with Wings Upsets Theories: Professor Has Given Ocular Proof of His Accomplishments–Tells of His Early Disappointments and Grit," *San Jose Mercury,* March 29, 1905. Montgomery is quoted extensively in the article.

4. John J. Montgomery, "Address by Prof. John J. Montgomery: Before the Aeronautical Society, April 21, 1910," *Engineers' List* 21, no. 1 (1912): 15–21. This lecture to the Aeronautic Society of New York was published posthumously.

5. Octave Chanute, *Progress in Flying Machines* (New York: American Engineer and Railroad Journal, 1894).

6. Although other authors (post-1940) have adopted 1883 as the year of the first flights with Montgomery's first fixed-wing glider, Montgomery himself clearly suggested that the glider was built in 1883 and stated explicitly that many successful flights were made with it in 1884. Primary statements on this were given at the time of Montgomery's first press conference (of March 25, 1905) and published in "Machine with Wings Upsets Theories"; "Years of Research Applied to Solving the Problem," *San Jose Mercury Evening News,* March 31, 1905; and "Third Tests Are All Successful," *San Francisco Bulletin,* March 26, 1905. Montgomery is quoted extensively in these articles.

7. Jane E. Montgomery, affidavit, February 3, 1945 [SCU-JJM]. The embellishments in the quotation are Jane's.

8. Historic climatic data for San Diego courtesy of Ted Mackechnie, San Diego Office, National Weather Service, San Diego, Calif.

9. Jane E. Montgomery affidavit. The embellishments were Jane's.

10. James P. Montgomery to Congressman Robert C. Wilson, June 5, 1944 [SCU-JJM].

11. Error in transcription. All other versions of this lecture (published and unpublished) state 600 feet, not 600 yards.

12. John J. Montgomery, "Address by Prof. John J. Montgomery." For James Montgomery's firsthand account of the first gliding flights, see James P. Montgomery, affidavit, December 29, 1944 [SCU-JJM].

13. "Montgomery Hits Wright's Patent: California Professor Claims He Invented Warping Wings Back in 1885," *New York World,* April 24, 1910. Regarding 600 feet as the maximum length of the first flights, see also James P. Montgomery, in answer to question 16, in Court (Equity No. 33852), January 13, 1919; Dennis J. Kavanagh, "The Story of the Aeroplane," in *The Aeroplane* (Santa Clara: Aeroplane Advertising Co., 1905), 16–23.

14. James P. Montgomery, in answer to cross-question 44, in Court (Equity No. 33852), January 13, 1919. On that same day of court testimony (January 13, 1919) James also stated in answer to question 15 that the longest glides accomplished with the craft of 1883–84 would glide up to 600 feet.

15. Jane E. Montgomery affidavit.

16. "Montgomery Hits Wright's Patent." Montgomery is quoted extensively in the article.

17. Cayley's coachman has sometimes been identified as "John Appleby, who would have been aged about 20 at the time of the flight."; J. A. D. Ackroyd, "Sir George Cayley, the Father of Aeronautics," pt. 2: "Cayley's Aeroplanes," *Notes and Records of the Royal Society of London* 56, no. 3 (2002): 347.

18. Kavanagh, "The Story of the Aeroplane."

Chapter 5. The Internal Work of the Wind

1. *San Diego Union,* April 14, 1884. See also the issues of May 24, 1884, June 4, 1884, and July 1, 1884.

2. Zach Montgomery, "Latter Day Reminiscences," *Family's Defender Magazine and Educational Review* 4, no. 1 (1884): 335–41.

3. *San Diego Union,* September 3, 1884. See also the issues of September 17, October 5, and October 28, 1884.

4. Assembly Bill 348 (the "Montgomery Bill").

5. *Santa Clara College Catalog, 1882–1887* (Santa Clara: College Steam Press, 1887). The catalog confirms the attendance of Richard J. Montgomery (August 5, 1883–June 2, 1884) and James P. Montgomery (August 5, 1884–June 2, 1885), after which James completed his baccalaureate education at Georgetown College in Washington, D.C. Both Richard and James received commendation for their studies in various subjects at Santa Clara College.

6. John J. Montgomery, "Our Tutors in the Art of Flying," *Aeronautics* (New York) 8, no. 9 (1915): 99–100. "The mountain regions" refers to the San Ysidro Peak, a prominent peak within the San Ysidro mountain range at the eastern end of Otay Valley in San Diego County.

7. John J. Montgomery, "The Origin of Warping: Professor Montgomery's Experiments," *Aeronautics* (London) 3, no. 5 (1910): 63–64.

8. Ibid.

9. Melvin D. Johnson, interviewed by Edgar Hastings, May 10, 1958, for the San Diego Historical Society Oral History Program; John V. Hettich, "Checking Up on the Montgomery Family," *San Ysidro Border Press,* May 20, 1950. Johnson's presence as a neighbor in the period is documented in the 1892 Land Ownership Book for San Diego County.

10. Helen Dare, "Prof. J. J. Montgomery–The Personality of the Man," *San Francisco Call,* May 7, 1905. Montgomery is quoted extensively in the article.

11. Rose Blanco Clemens to Arthur D. Spearman, August 6, 1857 [SCU-JJM].

12. "Gallant Journey Film Recalls Childhood of Miss Arnold of Deland," *Deland Sun News* (Florida), February 28, 1947. Isabel Arnold (a neighbor of the Montgomery's in the 1880s) is quoted extensively in the article.

13. This apparatus was described by Jane Montgomery as well. Jane E. Montgomery, affidavit, February 3, 1945 [SCU-JJM].

14. John J. Montgomery, "The Early Development of the Warping Principle" (unpublished ms. written sometime in mid-1909) [SCU-JJM]. The mechanism he talks about here provided a degree of lateral stability as well as a degree of pilot-operated roll control.

15. Charles Burroughs, affidavit, February 26, 1920, notarized by Agnes G. Nello [SCU-JJM].

16. Ibid.

17. Jane E. Montgomery affidavit. For Stokes's assistance with flights, see Edward Stokes to Richard J. Montgomery, December 18, 1911 [SCU-JJM].

18. Stokes to Richard J. Montgomery, December 18, 1911.

19. Mary Fidelis McCarthy, in answer to question 20, in Court (Equity No. 33852), January 14, 1919.

20. John J. Montgomery to Margaret H. Montgomery, December 22, 1885 [SCU-JJM].

21. *San Diego Union*, June 28, 1885.

22. Montgomery, "Our Tutors in the Art of Flying."

23. Richard J. Montgomery, in answer to questions 22, 23, and 24, in Court (Equity No. 33852), January 13, 1919.

24. Ibid.

25. Ibid.

26. Montgomery, "The Origin of Warping."

27. Jane E. Montgomery, in answer to question 19, in Court (Equity No. 33852), January 13, 1919.

28. Zach Montgomery to Richard J. Montgomery, August 6, 1885. Written on letterhead of the Department of the Interior, Office of the U.S. Attorney General, Washington, D.C.

29. Montgomery, "Our Tutors in the Art of Flying."

30. John J. Montgomery, "Discussion on the Various Papers on Soaring Flight," in *Proceedings of the Conference on Aerial Navigation, Held in Chicago, August 1, 2, 3 and 4, 1893*, edited by M. N. Forney (New York: American Engineer and Railroad Journal, 1894), 247–49.

31. Octave Chanute to Herman W. L. Moedebeck, October 16, 1895, Octave Chanute Papers, Library of Congress Manuscript Division, Washington, D.C. [hereafter LoC-OC].

32. Montgomery, "The Origin of Warping."

33. Montgomery, "Discussion on the Various Papers on Soaring Flight."

34. Dennis J. Kavanagh, "The Story of the Aeroplane," in *The Aeroplane* (Santa Clara: Aeroplane Advertising Co., 1905), 16–23.

35. Ibid.

36. John J. Montgomery, "Soaring Flight" (unpublished ms., 1894, 131 pp.) [SCU-JJM, photocopy of original]. The Inventory of the John J. Montgomery Collection (Archives and Special Collections, Santa Clara University) indicates that the original resides at the Smithsonian National Air and Space Museum, Archives Division.

37. Montgomery, "Discussion on the Various Papers on Soaring Flight."

38. Ibid.

39. John J. Montgomery to Margaret Helena Montgomery, December 23, 1885 [SCU-JJM].

40. Ibid.

41. Ibid., emphasis in the original.

42. Louis Mouillard to Octave Chanute, June 14, 1890. From *The Chanute-Mouillard Correspondence, April 16, 1890 to May 20, 1897: Being the Letters Exchanged between Octave Chanute, American Engineer, and Louis-Pierre Mouillard, French Author and Student of Bird Flight, Mainly on the Subject of Aeronautics*, translated by Eugene Moritz and Dr. M. Louise Kraus, edited by Juliette Bevo-Higgins (San Francisco: E. L. Sterne, 1962), transcribed and available online at http://invention.psychology.msstate.edu/i/Chanute/library/Chanute_Mouillard/1890.html.

43. Ibid.

44. The third flying machine was constructed during the winter of 1885–86, with experiments in 1886.

45. Montgomery, "The Origin of Warping." Montgomery notes in this article that while the craft was effective in terms of control, the very limited distance of flights obtained

showed that the airfoil was not an effective design. Chanute misstated the results of these experiments when he reported that "it was not successful; several trials were made with it but no effective lift could be obtained"; Octave Chanute, *Progress in Flying Machines* (New York: American Engineer and Railroad Journal, 1894), 249.

46. Montgomery's control device was similar in function to the "pitcheron" of Professor Albert. A. Merrill used in the 1910s and 1920s, and to the control surfaces used in *The Antoinette* during 1906–1909 and thereafter.

47. Winsor Josselyn, notes from unpublished interview with James P. Montgomery, Oakland, Calif., April 8, 1939 [SCU-JJM].

Chapter 6. The Path to Recognition

1. Zach Montgomery, *Poison Drops in the Federal Senate: The School Question from a Parental and Non-sectarian Stand-point* (Washington, D.C.: Gibson Bros., 1886). See also Zach Montgomery, *Bimetallism: Bondage or Blood–The Money Question Made Plain* (Los Angeles: Z. Montgomery and Sons, 1895).

2. Zach Montgomery, *Sapping the Foundation of Our Liberties! Remarks by Hon. Zach. Montgomery, before the Roman Catholic Sunday School Teachers, Sunday, July 6th, 1873* (San Francisco: Zach Montgomery, 1873).

3. *San Diego Union,* May 23, 1885.

4. *New York Times,* May 26, 1885. See also *New York Times,* May 29 and March 3, 1886; *Nation,* May 1885; *Atlanta Constitution,* May 26, 1885; *Chicago Times,* May 26, 1885; *New York Herald,* May 26, 1885.

5. *Journal of the Executive Proceedings of the Senate of the United States,* vol. 25: *March 3, 1885–March 4, 1887* (Washington, D.C.: U.S. Government Printing Office, 1901), 333, 350.

6. Ibid.

7. Zach Montgomery, *Bimetallism,* iv.

8. After a year of preparatory study at Santa Clara College in 1885–85, James completed his undergraduate studies at Georgetown University in Washington, D.C. He received his bachelor's of science (1888) and his bachelor's of law (1889) from Georgetown.

9. "Flying Machines," *New York Herald,* excerpted in *Worcester Daily Spy* (Mass.), November 19, 1886.

10. Samuel Pierpont Langley, "Story of Experiments in Mechanical Flight," in *The Aeronautical Annual: Devoted to the Encouragement of Experiment with Aerial Machines, and to the Advancement of the Science of Aerodynamics,* edited by James Means (Boston: W. B. Clarke & Co., 1897), 11–25.

11. John Montgomery to James Montgomery, September 7, 1886 [SCU-JJM]

12. *Buffalo Courier,* August 26, 1886.

13. Octave Chanute, *Progress in Flying Machines* (New York: American Engineering and Railroad Journal, 1894), 197–98.

14. Tom D. Crouch, *A Dream of Wings: Americans and the Airplane, 1875–1905* (Washington, D.C.: Smithsonian Institution Press, 1989), 40–41.

15. Octave Chanute to Louis-Pierre Mouillard, December 12, 1892 [LoC-OC].

16. Louis-Pierre Mouillard to Octave Chanute, January 9, 1893 [LoC-OC].

17. Octave Chanute to Louis-Pierre Mouillard, December 6, 1893 [LoC-OC]. By the time of the above-referenced letter, Chanute and Montgomery had met and discussed the Otay experiments.

18. Leander A. Redman, in answer to question 10, in Court (Equity No. 33852), January 16, 1919. Redmond used the spelling Leander in official documents but signed his name Lander as well.

19. Ibid.

20. Redman clarified in direct court testimony (1919) that his correspondence with Montgomery and others was among the personal belongings that were destroyed in the fires that consumed his place of residence following the Great San Francisco Earthquake of 1906.

21. Ibid.

22. Ibid.

23. Leander A. Redman, "California Entertaining a Genius Unawares," *San Francisco Bulletin*, March 27, 1905; reprinted in Redman, *Professor Montgomery's Discoveries in Celestial Mechanics* (San Francisco: Pernau-Walsh Printing Co., 1919), 5–6.

24. John V. Hettich, "Montgomery Chronology" (unpublished ms., ca. 1946), Ernest Jones Aeronautical Collection, Smithsonian National Air and Space Museum, Archives Division, Washington, D.C. [hereafter NASM-ELJ].

Chapter 7. Chicago, a Forum for the Outsider

1. Albert F. Zahm, "Octave Chanute: His Work and Influence in Aeronautics," *Scientific American* 104, no. 19 (1911): 463.

2. Howard L. Scamehorn, *Balloons to Jets: A Century of Aeronautics in Illinois, 1855–1955* (Carbondale: Southern Illinois University Press, 1957), 24.

3. Zahm, "Octave Chanute."

4. Albert F. Zahm to John Hettich, April 17, 1946, quoted in John Hettich to Ernest La Rue Jones, May 17, 1946 [NASM-ELJ].

5. Octave Chanute, "Sailing Flight," in *Aeronautical Annual: Devoted to the Encouragement of Experiment with Aerial Machines, and to the Advancement of the Science of Aerodynamics,* edited by James Means (Boston: W. B. Clarke & Co., 1894), 60–76, 98–127.

6. Octave Chanute to Louis P. Mouillard, June 17, 1892 [LoC-OC].

7. Ibid.

8. Ford A. Carpenter, *The Climate and Weather of San Diego, California* (San Diego: San Diego Chamber of Commerce, 1913).

9. Scamehorn, *Balloons to Jets,* 24.

10. Helen Dare, "Prof. J. J. Montgomery–The Personality of the Man," *San Francisco Call,* May 7, 1905.

11. John J. Montgomery to Mary C. Montgomery, August 25, 1893, written at the Engineering Headquarters of the 1893 Chicago World's Fair [SCU-JJM]. It is unclear whether those words of introduction were delivered to the audience by Chanute, Zahm, or Charles Bonney.

12. John J. Montgomery, "The Origin of Warping: Professor Montgomery's Experiments," *Aeronautics* (London) 3, no. 5 (1910): 63–64.

13. As corroborated by correspondence between Chanute and Montgomery in 1894, and between Chanute and Zahm during this same period, Montgomery's two lectures were recorded by a stenographer. Montgomery reprised this second lecture and retitled it "The Origin of Warping" in 1910 at the Aeronautic Society of New York (see chap. 15).

14. Zahm to Hettich, April 17, 1946.

15. Montgomery, "The Origin of Warping."

16. Winsor Josselyn, notes from personal interview with Regina Montgomery at the Press Club, San Francisco, March 21, 1946 [SCU-JJM]. The notes consist of brief quotes of her responses to Josselyn's questions.

17. Ibid.

18. Ibid.

19. Albert F. Zahm, diary for the Conference on Aerial Navigation, Chicago, entry for August 1, 1893: "Mr. Montgomery of Santa Clara College came and described his gliding experiments" [Albert Francis Zahm Papers, UND]. See also Zahm's biography of Montgomery in "Catholic Contributions in the Field of Aeronautics," in *Catholic Builders of the Nation* (Boston: Continental Press, 1923).

20. Octave Chanute to John Montgomery, August 8, 1895 [LoC-OC].

21. Montgomery, "The Origin of Warping."

22. *Los Angeles Times,* March 26, 1894.

23. Montgomery, "The Origin of Warping."

24. John J. Montgomery, "Discussion on the Various Papers on Soaring Flight," in *Proceedings of the Conference on Aerial Navigation, Held in Chicago, August 1, 2, 3 and 4, 1893,* edited by M. N. Forney (New York: American Engineer and Railroad Journal, 1894), 247–49.

25. Dare, "Prof. J. J. Montgomery." "I worked it all out 10 years ago" refers to the fact that in 1895 he worked out the theoretical considerations for the airfoils of his aeroplane design of 1903.

26. Dr. Mark Ardema, personal communication to Craig Harwood, 2004. Also John H. McMasters, email message to Craig Harwood and Gary Fogel, April 12, 2006; and Gary A. Flandro, "Montgomery's Aerodynamics" (unpublished ms., 2011, 8 pp.).

27. Flandro, "Montgomery's Aerodynamics." This manuscript was forwarded to Harwood and Fogel by its author in June 2011. Craig Harwood, personal collection.

28. Ibid.

29. McMasters, email message to Harwood and Fogel, April 12, 2006.

30. John D. Anderson, Jr., *A History of Aerodynamics and its Impact on Flying Machines* (Cambridge: Cambridge University Press, 1997), 246.

31. Flandro, "Montgomery's Aerodynamics."

32. Ibid. We will describe these various forms in subsequent chapters.

33. Octave Chanute to Albert F. Zahm, January 5, 1894, excerpted in John V. Hettich, "Montgomery Chronology" (unpublished ms., ca. 1946) [NASM-ELJ].

34. Margaret H. Montgomery, in answer to question 28, in Court (Equity No. 33852), January 13, 1919.

35. Curiously, although Chanute carefully preserved his correspondence, no surviving copies of John's letters to Chanute from 1893–95 are known to exist in any collection containing Chanute's correspondence.

36. Octave Chanute, *Progress in Flying Machines* (New York: American Engineer and Railroad Journal, 1894).

37. Tom D. Crouch, *A Dream of Wings: Americans and the Airplane, 1875–1905* (Washington, D.C.: Smithsonian Institution Press, 1989), 167.

38. John J. Montgomery to Mary C. Montgomery, August 25, 1893 [SCU-JJM].

Chapter 8. Other Pursuits

1. John J. Montgomery, "Petroleum Oven," German Patent 88,977, issued November 12, 1895.

2. The editors of *Scientific American* would later publish a very similar, but much more condensed, treatise by Montgomery after he had gone public (see chap. 11), titled "New Principles in Aerial Flight," *Scientific American Supplement* 1560 (November 25, 1905): 24991–93.

3. "Academy of Sciences: Professor Montgomery's Paper on the Flight of Birds," *Los Angeles Evening Express*, November 10, 1897.

4. *University of Santa Clara: A History from the Founding of Santa Clara Mission in 1777 to the Beginning of the University in 1912* (Santa Clara: University Press, 1912), 89.

5. Whereas today the term "electrician" is used in the narrow sense of "one who installs and/or maintains electrical equipment," at that time it was used to describe anyone who was engaged in investigations or expert work involving electrical phenomena. A more appropriate modern equivalent would be "electrical engineer" or "physicist."

6. "Testimony of John Joseph Montgomery in the Impeachment Hearing of Judge Lucas Flattery Smith," February 17, 1905, in "The Assembly of the State of California," vol. 1: "Testimony on Behalf of the Memorialists," 265, transcribed and indexed by Stanley D. Stevens, 2005–10. Collection of Charles Bruce Younger Jr., Hihn-Younger Archive, box 51, Special Collections, McHenry Library, University of California Santa Cruz, California. See also John J. Montgomery to Editor, *San Jose Mercury and Herald,* January 2, 1905.

7. "A Giant among Giants," in Francis J. Weber, *Encyclopedia of California's Catholic Heritage, 1769–1999* (Mission Hills, Calif.: Saint Francis Historical Society, 2000), 855.

8. Oliver Miles, "Review of the Life of the Late Hon. Zach Montgomery," *Tidings* 6, no. 40 (October 6, 1900), 1–4 (a paper read before the Newman Club, Los Angeles, September 28, 1900).

9. "Santa Clara College Observatory Is Doing Good Astronomical Work," *San Francisco Call,* October 17, 1903.

10. "An American Passion Play," editorial, *New York Times,* June 30, 1901. See also "Passion Play Scores Great Triumph," *San Francisco Call,* May 26, 1903.

11. Santa Clara College would not achieve status as an accredited university until 1912.

12. "Local Scientists Invent a New System of Wireless Telegraphy," *San Francisco Call,* March 6, 1904.

13. *San Francisco Monitor,* clipping from sometime in the spring of 1903, excerpted in John V. Hettich, "Montgomery Chronology" (unpublished ms., ca. 1946) [NASM-ELJ].

14. Montgomery to Editor, January 2, 1905. Guglielmo Marconi, the famous Italian pioneer of wireless telegraphy, was impressed with the Montgomery-Bell telegraph when he visited Santa Clara College in 1933.

15. "Chanute Deplores Wright Patent Suits: Early Experimenter Who Turned His Data Over to Them Thinks They Will Check Progress," *New York Times,* January 23, 1910. See also "Dr. Chanute Denies Wright Flying Claim," *New York World,* January 17, 1910. By "my biplane" he meant the Chanute biplane glider that he had developed in the mid-1890s

16. "Dr. Chanute Denies Wright Flying Claim," *New York World,* January 17, 1910.

Chapter 9. A California Impetus

1. Howard Lee Scamehorn, "Thomas Scott Baldwin: The Columbus of the Air," *Journal of the Illinois State Historical Society* 49, no. 2 (1956): 163–89.

2. Central Park was later renamed Golden Gate Park.

3. "Battle of the Airship Men Goes Merrily On and On and Now's in Court," *San Francisco Examiner,* June 2, 1905. August Greth is quoted extensively in the article, recounting what he had been told by Van Tassel about the Baldwin/Van Tassel collaboration.

4. "Falling 1,000 Feet: It Was with a Parachute and the Man Touched Bottom Safely," *New York Sun,* February 2, 1887. See also "Dropped 1,000 Feet From a Balloon," *Fort Worth Gazette,* February 5, 1887, and "Falling 1,000 Feet," *New York Sun,* February 18, 1887, 1, col. 6.

5. "That Big Jump," *Los Angeles Times,* July 27, 1888.

6. Ibid.

7. Ibid.

8. In 1913–14, Broderick was influential in convincing the U.S. Army of the potential of parachute jumps from airplanes.

9. Baldwin held a British patent for "Improvements in Parachutes" in 1888 (U.S. Patent No. 10,937). He is often credited in the literature as the inventor of the collapsible silk parachute.

10. "Fatalities from Parachute Jumping," *Philadelphia Enquirer,* September 29, 1901.

11. "Veteran of 800 Flights to Run First of Government Aerial Fleet," *Fort Wayne Sentinel,* March 21, 1908. The article is based on an interview with Baldwin.

12. "Greth, August," in John William Leonard and Albert Nelson Marquis, *Who's Who in America* (Chicago: Marquis Who's Who, 1906), 4:726.

13. "Aeronaut Building Airship at San Jose; Noted Parachute Gymnast Makes Design," *San Jose Mercury News (Evening),* October 20, 1902.

14. "Dirigible Balloon Is a Success," *San Francisco Call,* April 24, 1904.

15. "Dr. Greth Rides a Ship; High above the Earth," *Omaha World Sunday Herald,* April 24, 1904.

16. "Fell in the Water: Successful Voyage of the Greth Airship," *San Jose Mercury News,* October 19, 1903.

17. "Airship Builder Has Begun Two Lawsuits: Montgomery Seeks to Restrain Baldwin and Seeks Damages," *San Jose Mercury (Evening) News,* April 7, 1905. Based on an interview with Montgomery.

18. "War of Airships Comes Nigh to Costing an Aeronaut His Life," *San Francisco Examiner,* May 26, 1905. Montgomery is quoted extensively in the article.

19. John J. Montgomery to James P. Montgomery, May 15, 1904 [James P. Montgomery Papers, SCU-JJM].

20. Charles Harvard Gibbs-Smith, *Sir George Cayley (1773–1857)* (London: HMSO, 1948), 149.

21. Thomas S. Baldwin to John J. Montgomery, June 20, 1903 [SCU-JJM].

22. John J. Montgomery to Octave Chanute, April 11, 1905 [LoC-OC].

23. "Presentation of Langley Medal to Messrs. Wilbur and Orville Wright: Historical Address by Dr. Alexander Graham Bell," in "Report of Executive Committee and Proceedings of Board of Regents, Year Ending June 30, 1910" (Smithsonian Publication 2001), reprinted in *Annual Report of the Board of Regents of the Smithsonian Institution . . . for the Year Ending June 30, 1910* (Washington, D.C.: GPO, 1911), 104–108; also *Science* 31, no. 792 (1910): 336.

24. "Aeronaut's Success: Dr. Greth Makes an Ascent in Airship; Will Contest for World's Fair Prize with Santos-Dumont," *Biloxi Daily Herald,* August 12, 1903.

25. John J. Montgomery, "The Origin of Warping: Professor Montgomery's Experiments," *Aeronautics* (London) 3, no. 6 (1910): 63–64.

26. John H. Leonard, in answer to questions 11 and 12, in Court (Equity No. 33852), January 25, 1919.

27. John J. Montgomery, interview at Santa Clara College, March 25, 1905, quoted in "Careful Research of Several Years Assured Success," *San Francisco Bulletin,* March 26, 1905.

28. Victor Lougheed, *Aeroplane Designing for Amateurs* (Chicago: Reilly & Britton Co., 1912), 21.

29. Thomas S. Baldwin to John J. Montgomery, November 23, 1903 [SCU-JJM]. In this letter Baldwin acknowledged receipt of a letter from Montgomery, and acknowledged the success of the models that Montgomery described.

30. Ibid.

31. "War of Airships Comes Nigh."

32. Orville Wright, journal entry for December 17, 1903, Library of Congress American Memory Collection: The Wilbur and Orville Wright Papers, http://memory.loc.gov/ammem/wrighthtml/wrighthome.html [hereafter LoC-WOW].

33. Orville Wright, "The Wright Brothers' Aeroplane," *Century Magazine* 76 (September 1908): 641–50.

34. John J. Montgomery to James P. Montgomery, May 15, 1904.

35. Ibid.

36. The term "warping," or "wing warping," was common by the time of John's 1910 lecture from which the quote comes.

37. John J. Montgomery to Ellen Montgomery, December 15, 1903 [SCU-JJM].

38. Contract between J. J. Montgomery of the County of Santa Clara, State of California, and T. S. Baldwin of the City and County of San Francisco, State of California, April 28, 1904 [SCU-JJM].

39. "War of Airships Comes Nigh."

40. Contract between Montgomery and Baldwin.

41. Ibid.

Chapter 10. A Rude Awakening

1. "Dirigible Balloon Is a Success," *San Francisco Call,* April 24, 1904.

2. Thomas S. Baldwin, "I Will Fly Over and Under the Brooklyn Bridge," *World Magazine* (supplement to the *New York World*), November 27, 1904.

3. Thomas S. Baldwin, "The High Seas of Space," *National Magazine* 28, no. 4 (July 1908): 457–60. *The California Arrow* is discussed later in this chapter.

4. "Inventor of New Airship Has Trouble with Baldwin," *San Jose Mercury,* April 7, 1905. Montgomery is quoted extensively in the article.

5. Ibid.

6. John J. Montgomery to Octave Chanute, April 11, 1905 [LoC-OC].

7. John J. Montgomery, interview at Santa Clara College, March 25, 1905, quoted in "Careful Research of Several Years Assured Success," *San Francisco Bulletin,* March 26, 1905.

8. John J. Montgomery, "The Origin of Warping: Professor Montgomery's Experiments," *Aeronautics* (London) 3, no. 5 (1910): 63–64.

9. "War of Airships Comes Nigh to Costing an Aeronaut His Life," *San Francisco Examiner,* May 26, 1905.

10. "Says Baldwin Borrowed Ideas: Prof. J. J. Montgomery of San Jose Claims the Credit for Airship–Will Go to Law," *San Francisco Chronicle,* December 27, 1904. Montgomery is quoted extensively in the article.

11. "Inventor of New Airship Has Trouble with Baldwin."

12. "War of Airships Comes Nigh."

13. "Newly Constructed Airship Will Fly over San Jose," *San Jose Mercury Evening,* July 30, 1904.

14. J. Mayne Baltimore, "The New Baldwin Airship," *Scientific American* 91, no. 9 (August 27, 1904): 147.

15. Kenneth M. Johnson, *Aerial California: An Account of Early Flight in Northern and Southern California, 1849 to World War I* (Los Angeles: Dawson's Book Shop, 1961), 27.

16. "The French Aeroplane Contest," *Scientific American* 92, no. 11 (March 18, 1905): 220. Later in 1922, French pilot Alexis Maneyrol set an international record when he soared for 3 hours, 21 minutes, 7 seconds in Peyret's *Tandem Monoplane* (also known as *The Alérion*) to win an international glider meet at Ilford, Sussex, England.

17. "Aeronautics," *Automotor Journal* 9, no. 43 (October 22, 1904): 1243–44. Alvarez is consistently referred to as "Señor Alvarez" in press reports.

18. Ibid.

19. Glenn H. Curtiss and Augustus Post, *The Curtiss Aviation Book* (New York: Frederick A. Stokes Co., 1912), 30.

20. "Question of Compensation Sends Baldwin from the Fair," *San Francisco Call,* November 11, 1904.

21. Thomas S. Baldwin, U.S. Patent No. 851,481, "Air-ship," filed November 21, 1904, allowed April 23, 1907.

22. "War of Airships Comes Nigh."

23. Baldwin, "I Will Fly."

24. "Inventor of New Airship Has Trouble with Baldwin."

25. "Airship Inventor Visits San Jose: Captain Baldwin, Builder of the California Arrow, Talks of His Plans and Future of Aerial Navigation," *San Jose Daily Mercury*, December 5, 1904.

26. "Father Bell and Captain Baldwin: Priest-Scientist of Santa Clara College Gives Credit for Success of Aerial Experiments to Professor Montgomery," *San Jose Daily Mercury*, December 6, 1904. Bell is quoted extensively in the article.

27. "The Aeronautic Hub," *Los Angeles Herald*, March 7, 1905.

28. "Aerial Experts Organize Club," *Los Angeles Herald*, August 27, 1905.

29. "Airship Race Not a Success," *San Francisco Call*, September 11, 1905.

30. "Did Baldwin Have Ideas for Airship?" *San Jose Mercury News*, December 27, 1904.

Chapter 11. Going Public

1. Jane E. Montgomery, in answer to question 110, in Court (Equity No. 33852), January 13, 1919.

2. Ibid.

3. "Montgomery's New Airship," *San Jose Evening News*, March 8, 1905; "New Airship Will Fly Monday," *San Jose Evening News*, March 11, 1905; "Airship Flies Like a Bird," *San Jose Evening News*, March 18, 1905.

4. John J. Montgomery, interview at Santa Clara College, March 25, 1905, quoted in "Careful Research of Several Years Assured Success," *San Francisco Bulletin*, March 26, 1905. Montgomery is quoted extensively in the article.

5. John J. Montgomery to Octave Chanute, April 11, 1905 [LoC-OC]. See also "Careful Research of Several Years Assured Success."

6. "Careful Research of Several Years Assured Success." The flight of March 20, 1905, is also recounted by Father Richard Bell in "The Montgomery Aeroplane," *San Francisco Bulletin*, March 25, 1905; and "Flying May Soon Be Easy for Everyone," *Saturday Bee* (Sacramento), April 1, 1905. For Maloney's account of this flight, see "Daring Aeronaut Tells How He Sailed in Mid-Air on Most Remarkable Airship Ever Invented," *Bulletin* (San Francisco), March 26, 1905.

7. "Daring Aeronaut Tells How He Sailed in Mid-Air."

8. Amos I. Root to the Wright Brothers, March 22, 1905 [LoC-WOW].

9. "Careful Research of Several Years Assured Success." See also "Machine with Wings Upsets New Theories," *San Jose Mercury*, March 29, 1905. Both articles are based on an interview conducted with Montgomery at a March 25, 1905, press conference at Santa Clara College.

10. "Years of Research Applied to Solving the Aerial Problem," *San Jose Evening Mercury*, March 31, 1905. Montgomery is quoted extensively in the article.

11. Ibid.

12. "Aerial Flight Nearly Achieved," *Washington Post*, March 30, 1905.

13. "War of Airships Comes Nigh to Costing an Aeronaut His Life," *San Francisco Examiner*, May 26, 1905.

14. Octave Chanute to Wilbur Wright, April 4, 1905 [LoC-OC].

15. Octave Chanute to John J. Montgomery, April 4, 1905 [SCU-JJM].

16. Octave Chanute to Maj. Hermann Moedebeck, April 5, 1905 [LoC-OC].

17. Ibid.

18. Octave Chanute to Wilbur Wright, April 16, 1905 [LoC-OC].

19. Two California-based newspapers mentioned the Wrights' work prior to 1905: the *Los Angeles Times,* December 19, 1903, and the *San Francisco Call,* January 8, 1904, and October 9, 1904. There may be others as well.

20. Amos I. Root to Wilbur Wright, April 17, 1905 [LoC-WOW].

21. Amos I. Root, "Flying Machines Up to Date: Wright Brothers Still Ahead," clipping from *Gleanings in Bee Culture,* dated by hand "April 1, 1905" [LoC-WOW].

22. Amos I. Root to Wilbur Wright, April 17, 1905 [LoC-WOW].

23. Bill of Complaint in the Superior Court of the State of California and the County of Santa Clara, dated April 7, 1905; plaintiff: John J. Montgomery; defendant: Thomas S. Baldwin [SCU-JJM]. See also "War of Airships Comes Nigh."

24. Octave Chanute to John J. Montgomery, April 16, 1905 [SCU-JJM].

25. Ibid.

26. Octave Chanute to John J. Montgomery, April 28, 1905 [SCU-JJM].

27. Octave Chanute to Wilbur Wright, April 28, 1905 [LoC-WOW].

28. John J. Montgomery to Ellen Montgomery, March 21, 1905 [SCU-JJM].

29. John J. Montgomery to James P. Montgomery, May 1, 1905 [SCU-JJM].

30. John J. Montgomery to Octave Chanute, April 30, 1905 [LoC-OC].

31. John J. Montgomery to James P. Montgomery, May 1, 1905 [SCU-JJM].

32. Ibid.

33. "Aeroplane Soars Downward Like Graceful Bird," *San Francisco Call,* April 30, 1905. Richard Bell is quoted extensively in the article.

34. "First Man to Fly Described the Sensation," *San Francisco Chronicle,* April 30, 1905. Daniel Maloney is quoted extensively in the article.

35. Dennis J. Kavanagh, "The Montgomery Aeroplane: Details of the California Air Craft Which Is Attracting World-Wide Attention," *Popular Mechanics* 7, no. 7 (July 1905): 703–707.

36. Ibid. The original citation for Bell's remark remains unknown, but a slightly shorter version of the quote (probably paraphrased) initially appeared in an article by Kavanagh titled "The Flight of April 29th," in the May 1905 issue of the Santa Clara College magazine, *The Redwood.* The article was reprinted in *The Aeroplane* (Santa Clara: Aeroplane Advertising Co., 1905), 27–30. The slightly shorter quote appears as "all subsequent attempts in aviation must begin with the Montgomery Machine."

37. See Charles B. Hayward, "The Montgomery Aeroplane," *Scientific American* 92, no. 20 (1905): 404. See also "Flying-Machine Plunges into Space 4,000 Feet above the Earth: Towed Up by a Balloon and Cut-Loose–Most Daring Test Ever Made," *Popular Mechanics* 7, no. 6 (1905): 613; Kavanagh, "The Montgomery Aeroplane"; "Aeronautics," *Automotor Journal* 10 (1905): 1079; Carl Dienstbach, "Ein Flug von 20 Minuten mit dem Gleitapparat von Montgomery," *Illustrierte aëronautische Mitteilungen* 9, no. 8 (1905): 255–58; P. DeMeriel, "Un aeroplane a 1200 meters," *La Nature,* November 25, 1905, 412; Hermann Moedebeck, *Fliegende Menschen! Das Ringen um die Beherrschung der Luft mittels Flugmaschinen* (Berlin: O. Salle, 1905); "Aeronautics," *Messenger* 44, no. 1 (1905): 102; "Notizen," *Wiener Luftschiffer-Zeitung* 4, no. 8 (1905): 169–70; Edward J. Wheeler, "An Aeroplane That Actually Soars," *Current Literature* 39 (1905): 184–86; Henri Coupin,

"Un descende de 1200 metres en aeroplane," *Le magasin pittoresque,* March 1905, 139–40.

38. Albert F. Zahm, *Aerial Navigation: A Popular Treatise on the Growth of Air Craft and on Aeronautical Meteorology* (New York: D. Appleton and Co., 1911), 252.

39. "Aeroplane Soars Downward Like Graceful Bird."

40. Kavanagh, "The Flight of April 29th."

Chapter 12. Tragedy

1. Octave Chanute to Wilhelm Kress, June 5, 1905 [LoC-OC].

2. Octave Chanute to Samuel P. Langley, May 25, 1905 [LoC-OC].

3. To avoid the frequent mispronunciation of his surname, Victor adopted the more phonetically correct spelling Lougheed. Malcolm and Allan retained the original spelling of Loughead until sometime in the 1920s, when for business reasons they adopted the spelling Lockheed.

4. Richard J. Montgomery, in answer to question 16, in Court (Equity No. 33852), January 17, 1919.

5. Ibid.

6. Ibid.

7. "Drops Safely from the Sky: Accident to Professor Montgomery's Aeroplane Shows the Merits of the Machine," *San Francisco Call,* May 22, 1905. See also "War of Airship Comes Nigh to Costing Aeronaut Life; Tampering Is Done," *San Francisco Examiner,* May 26, 1905.

8. "Who Saw the Airship This Morning? California Arrow Was Reported to Have Flown over San Jose," *San Jose Evening News,* May 13, 1905.

9. Jane E. Montgomery, in response to question 96, in Court (Equity No. 33852), January 13, 1919.

10. "Baldwin Sues Rival for Libel: Inventor of Aeroplane Will Fight Aeronaut Who Brings Accusation against Him," *Boston Journal,* July 10, 1905.

11. "They're Lies, Declares Baldwin," *San Francisco Examiner,* June 1, 1905. Baldwin is quoted extensively in the article.

12. Ibid.

13. Ibid.

14. Waldon Fawcett, "Captain T.S. Baldwin's Dirigible Balloon," *Automobile* 8, no. 18 (1905): 487–88.

15. "Battle of the Airship Men Goes Merrily On and On and Now's in Court," *San Francisco Examiner,* June 2, 1905.

16. Ibid.

17. "Inventor of Aeroplane Submits to Arrest: Professor Montgomery Is Charged with Libel by Aeronaut Baldwin," *San Jose Mercury News,* June 2, 1905.

18. John H. Pierce, "Aeroplane Did Not Sail in the Air: Balloon Is Wrecked as It Is About to Carry Airship from Ground." *Oakland Inquirer,* June 26, 1905.

19. "Flight of Aeroplane and Airship Yesterday: Prof. Montgomery's 'Santa Clara' and Heaton's 'Messenger' Do Well," *San Jose Mercury and Herald,* July 16, 1905.

20. Roy D. Graves, open letter, "Personal statements regarding Montgomery's death," December 1958 [SCU-JJM]. See also Rev. Charles E. Leahy, S.J., affidavit, July 7, 1962 [SCU-JJM].

21. John J. Montgomery to Octave Chanute, November 12, 1905 [LoC-OC].

22. Ibid. See also Richard H. Bell, S.J., to the editor, *Messenger* 45, no. 3 (September 1905): 321.

23. Montgomery to Chanute, November 12, 1905.

24. Leahy affidavit, July 7, 1962. Leahy, who later became a prominent priest in the San Francisco Bay area, had been hired to clean John's workshop/laboratory in the period 1902–1905.

25. Graves, open letter, December 1958.

26. Ibid.

27. Arthur Dunning Spearman, *John Joseph Montgomery, 1858–1911: Father of Basic Flying* (Santa Clara: University of Santa Clara, 1967), 86.

28. Maloney is interred at Santa Clara Mission Cemetery, Santa Clara, Calif.

29. "Killed: Brave Aeronaut Dan Maloney; Drops 2,000 Feet to His Death," *Santa Clara News,* July 18, 1905.

30. Testimony of H. J. Sheven in "Aeronaut's Daring May Have Caused Fall," *San Jose Mercury News,* July 19, 1905. See also testimony of Charles A. Nash, Robert Coward, and Reverend Richard H. Bell, S.J., at coroner's inquest on death of Daniel Maloney. The location of the original coroner's inquest remains unknown. All statements from the above witness were quoted in the above article. Portions of *The Santa Clara* remain in the collection of the National Air and Space Museum in Washington, D.C.

31. "Aeroplane Fell 3,000 Feet," *New York Times,* July 19, 1905.

32. Albert F. Zahm, "The Undeveloped Art of Soaring: Its Theory and Practice," *Scientific American,* Supplement 1872 (November 18, 1911): 332.

33. Wilbur Wright to Octave Chanute, August 6, 1905 [LoC-OC].

34. Octave Chanute to Wilbur Wright, September 6, 1905 [LoC-OC].

35. Wright to Chanute, August 6, 1905.

36. Octave Chanute, "Montgomery," in *Pocket-Book of Aeronautics,* edited by Hermann W. L. Moedebeck, translated by W. Mansergh Varley (London: Whittaker & Co., 1907), 309–10.

37. "Libel Proceedings Dismissed," *Los Angeles Herald,* July 19, 1905. See also "Libel Case Is Dismissed," *San Francisco Call,* July 19, 1905.

Chapter 13. Wings in Tandem

1. Richard Montgomery's home in Oakland served as the new family "headquarters" in 1905.

2. C. E. Warren to John J. Montgomery, July 20, 1905 [SCU-JJM].

3. "Two Airships under Construction," *San Jose Mercury News,* August 16, 1905.

4. Leo Ruth, Sr., undated open letter [SCU-JJM].

5. "Two Aeroplanes under Construction."

6. "Invited to Milan," *San Jose Evening News,* August 31, 1905.

7. "California Has a Modern Darius Green," *Kansas City Star,* September 3, 1905.

8. We will refer to this new craft as *The Defolco. The California* remained as before and was retired from use.

9. "Montgomery's New Aeroplane May Make Flight at Sacramento," *San Jose Mercury,* August 31, 1905.

10. "The Wonderful Aeroplane Will Sail To-Morrow Morning: Everything Is Satisfactory to Its Inventor, Prof. John J. Montgomery, of Santa Clara College," *Sacramento Bee,* September 6, 1905.

11. "Satisfactory Test Yesterday Afternoon of the Montgomery Aeroplane," *Sacramento Evening Bee,* September 9, 1905.

12. "Death Is the Helmsman," *Oakland Tribune,* July 19, 1905.

13. "Runaway Airship," *San Jose Mercury Evening News,* September 12, 1905.

14. Tom D. Crouch, *The Bishop's Boys: A Life of Wilbur and Orville Wright* (New York: W. W. Norton & Co., 1989), 297.

15. Ibid.

16. "Presentation of Langley Medal to Messrs. Wilbur and Orville Wright: Historical Address by Dr. Alexander Graham Bell," in "Report of Executive Committee and Proceedings of Board of Regents, Year Ending June 30, 1910" (Smithsonian Publication 2001), reprinted in *Annual Report of the Board of Regents of the Smithsonian Institution . . . for the Year Ending June 30, 1910* (Washington, D.C.: GPO, 1911), 104–108; also *Science* 31, no. 792 (1910): 336.

17. Edwin Black, *Internal Combustion: How Corporations and Governments Addicted the World to Oil and Derailed the Alternatives* (Washington, D.C.: Dialog Press, 2008), 85.

18. William W. Townsend (Patent Examiner, United States Patent Office) to Harry Toulmin (attorney for the Wright Brothers), July 14, 1903 [LoC-WOW]. See also F. I. Allen (Commissioner of Patents, United States Patent Office) to Harry Toulmin, November 8, 1904 [LoC-WOW].

19. Harry Toulmin to F. I. Allen, August 17, 1905 [LoC-WOW].

20. Ibid.

21. Legal Counsel for Plaintiffs (Hervey S. Knight et al.), in *Regina C. Montgomery et al. v. United States* (Equity No. 33852), Plaintiff's Request for Findings of Fact and Brief (Washington, D.C.: Washington Service Bureau, 1923) [SCU-JJM].

22. William W. Townsend to Harry Toulmin, December 2, 1905 [LoC-WOW].

23. Harry Toulmin to F. I. Allen, December 5, 1905 [LoC-WOW].

24. Wilbur Wright to Harry Toulmin, December 23, 1909 [LoC-WOW].

25. "Boy Aeronaut to Sail in an Aeroplane," *San Jose Evening News,* November 15, 1905.

26. Ibid.

27. "Machine Cannot Fly," *Oakland Tribune,* December 22, 1905.

28. Ibid.

29. John J. Montgomery to Octave Chanute, November 30, 1905 [SCU-JJM].

30. Augustus Post to John J. Montgomery, December 22, 1905. The original letter did not survive, but it was printed in full in "Montgomery Asked to Join Aero Club," *San Jose Sunday Mercury and Herald,* January 21, 1906, 7, col. 3.

31. David F. Wilkie to Congressman Bob Wilson, October 13, 1961 [SCU-JJM].

32. John J. Montgomery to Octave Chanute, May 27, 1906 [LoC-OC]. These flight trials are also described in Wilkie to Wilson, October 13, 1961.

33. Octave Chanute to John J. Montgomery, January 14, 1906 [SCU-JJM].

34. Octave Chanute to John J. Montgomery, February 16, 1906 [SCU-JJM].

35. "Aeronaut Dreams of Accident Which Occurs as in a Vision," *San Francisco Call,* February 23, 1906.

36. Montgomery to Chanute, May 27, 1906.

37. Ferdinand E. Blanchard, in answer to question 11, in Court (Equity No. 33852), January 25, 1919. For Wilkie's firsthand description of this flight, see David F. Wilkie, in answer to question 73, in Court (Equity No. 33852), January 23, 1919. The "barrel rolls" of the Montgomery aeroplane are also alluded to in "Aeronaut Dreams of Accident." See also "Menaced by Death," *Oakland Tribune,* March 19, 1906. The barrel rolls were made while the craft was in a steep angle of descent.

38. David F. Wilkie, in answer to question 95, in Court (Equity No. 33852), January 21, 1919.

39. Montgomery to Chanute, May 27, 1906.

40. "Aeroplanes, Balloons, etc.," *Washington Post,* February 25, 1906.

41. "Presentation of Langley Medal to Messrs. Wilbur and Orville Wright."

42. "Summary of Subjects Considered: Report of Board of Ordnance," in *Annual Reports of the War Department for the Fiscal Year Ended June, 30, 1906* (Washington, D.C.: Government Printing Office, 1906). The board considered the topic of "Balloons and Airships Submitted by RJ Montgomery," 252.

43. General Allen's reply to the letter of Knowland quoted in Richard J. Montgomery to Harry F. Stabler, May 28, 1921, John J. Montgomery Biographical File, Smithsonian National Air and Space Museum, Archives Division, Washington, D.C. [hereafter NASM-JJM]. Richard J. Montgomery to William Randolph Hearst, March 29, 1930 quoted in John V. Hettich, "Montgomery Chronology" (unpublished ms., ca. 1946) [NASM-ELJ].

44. Richard J. Montgomery to Harry F. Stabler, May 28, 1921.

45. Ibid.

46. "Plaintiff's Request for Findings of Fact and Brief" (Equity No. 33852) [SCU-JJM].

47. Harry Toulmin to Wilbur Wright, April 19, 1907 [LoC-WOW]. See also Wilbur Wright to Harry Toulmin, April 25, 1907 [LoC-WOW].

48. Montgomery to Chanute, May 27, 1906.

49. John J. Montgomery, "Some Early Gliding Experiments In America," *Aeronautics* (New York) 4, no. 1 (1909): 47–50.

50. Ibid.

Chapter 14. Baptism by Fire

1. Arthur J. Rodgers, "Re-creating the 1906 San Francisco Earthquake," *Science and Technology Review,* January–February 2007, https://www.llnl.gov/str/Sep06/Rodgers .html.

2. John J. Montgomery to Octave Chanute, May 27, 1906 [SCU-JJM].

3. There are no known references to the identity of these investors.

4. Richard J. Montgomery, in answer to question 129, in Court (Equity No. 33852), January 14, 1919.

5. Francis J. Weber, *George Thomas Montgomery, California Churchman* (Los Angeles: Westernlore Press, 1966), 33.

6. "Beloved Prelate at Rest in City of the Dead," *San Francisco Call*, January 15, 1907.

7. "Lost All His Balloons: Baldwin, the Aeronaut, Has an Airship East, However," *New York Times*, April 21, 1906.

8. Glenn H. Curtiss and Augustus Post, *The Curtiss Aviation Book* (New York: Frederick A. Stokes Co., New York, 1912), 30.

9. "The Wind Wagon," *Pearson's Magazine* 16, no. 5 (November 1906): 598.

10. Cornelius Reinhardt, affidavit, November 8, 1957 [SCU-JJM].

11. Ibid.

12. Ibid.

13. Victor Lougheed, *Aeroplane Designing for Amateurs* (Chicago: Reilly & Britton Co., 1912), 37–38.

14. Ibid.

15. John J. Montgomery, "The Mechanical Principles Involved in Soaring, Abstract to the Southern California Academy of Sciences, Los Angeles, November 9, 1897," *Los Angeles Evening Herald*, November 10, 1897; John J. Montgomery, "The Aeroplane: A Scientific Study," *The Redwood* 4, no. 5 (1905): 440–53, reprinted in *The Aeroplane* (Santa Clara: Aeroplane Advertising Company Co., 1905), 4–15; John J. Montgomery, "New Principles in Aerial Flight," *Scientific American Supplement* 1560 (1905): 24991–93.

16. John J. Montgomery, "Principles Involved in the Formation of Wing Surfaces and the Phenomenon of Soaring," *Aeronautics* (New York) 3, no. 4 (1908): 30–33; 3, no. 5 (1908): 39–40; 3, no. 6 (1908): 32–36, 4, no. 1 (1909): 43–46.

17. Ernest L. Jones, obituary for Cleve T. Shaffer, *CHIRP* (The Early Birds of Aviation), no. 71 (December 1964). An online version exists at http://earlyaviators.com/eshaffer.htm.

18. "Queen of the Pacific Wins Aerial Race," *San Francisco Call*, November 1, 1909.

19. Jack Carpenter, *Pendulum II: The Story of America's Three Aviation Pioneers–Wilbur Wright, Orville Wright, and Glenn Curtiss* (San Juan Capistrano: Arsdalen, Bosch & Co., 2003), 147.

20. James Allen, Office Memorandum–War Department, Office of the Chief Signal Officer, August 1, 1907 [LoC-WOW].

21. James Allen, "Signal Corps, U.S. Army Tentative Agreement for a Heavier Than Air Flying Machine," Office of the Chief Signal Officer, December 7, 1907 [LoC-WOW]. In response to the Wrights' insertions, Allen's letter of response (dated December 26, 1907) stated in part: "Your letter of the 18th instant with comments concerning the tentative specification for a flying machine, is appreciated, and was of value to this office in preparing the final specification."

22. Successful flights were made using this propeller arrangement on *The California Arrow II* at Hammondsport, New York, in June 1907.

23. "Veteran of 800 Flights to Run First of Government Fleet," *Fort Wayne Sentinel*, March 21, 1908.

24. Ibid.

25. Thomas S. Baldwin, "The High Seas of Space," *National Magazine* 28, no. 4 (July 1908): 457–60.

26. Fred Howard, *Wilbur and Orville: A Biography of the Wright Brothers* (New York: Ballantine Books, 1987), 302–303.

27. Octave Chanute to George A. Spratt, April 12, 1909 [LoC-OC].

28. "Visits Santa Clara College Observatory," *San Jose Mercury News*, December 12, 1908.

29. "A History of the University of Santa Clara Seismological Station" (undated ms.) [SCU-JJM].

30. "Our Tutors in the Art of Flying" was published posthumously in *Aeronautics* 8, no. 9 (1915): 99–100.

31. Kenneth M. Johnson, *Aerial California: An Account of Early Flight in Northern and Southern California, 1849 to World War I* (Los Angeles: Dawson's Book Shop, 1961), 66–67.

32. Claudia M. Oakes, *United States Women in Aviation through World War I* (Washington, D.C.: Smithsonian Institution Press, 1978), 11, 13.

33. Ibid.

34. "Montgomery First to Conquer the Air: Austrian Officials after Inquiry Give Palm to California Inventor," *San Francisco Examiner*, May 16, 1909. See also "Conquering the Air," *San Francisco Monitor*, June 12, 1909."Device to Charge Batteries,"

35. "Conquering the Air."

36. John J. Montgomery, U.S. Patent No. 947,141, "Rectifying Electric Currents," filed August 28, 1909, allowed November 1, 1910; and U.S. Patent No. 974,415, "Process for Compelling Electric Motors to Keep in Step with the Waves or Impulses of the Current Driving Them, and a Motor Embodying the Process," filed August 28, 1909, allowed November 1, 1910.

37. Lougheed held the following U.S. patents in aeronautics: No. 1,470,147 (1923, airfoil), No. 1,922,311 (1933, airfoil), No. 1,973,573 (1934, propeller), No. 2,175,204 (1939, propeller), No. 2,202,014 (1940, propeller).

38. Victor Lougheed, U.S. Patent No. 1,470,047, for "Wing for Flying Machines," filed February 4, 1919, allowed October 9, 1923.

39. Victor Lougheed, in answer to question 110, in Court (Equity No. 33852), January 28, 1919. Although Lougheed did not specify the model, it was most probably the *Loughead Model G.*

40. Victor Lougheed, in answer to question 3, in Court (Equity No. 33852), January 28, 1919.

41. Ibid.

42. Octave Chanute to Emerson Newell, October 14, 1909 [LoC-OC].

43. "Montgomery Aids Fight on Wrights by a New Action," *San Francisco Examiner*, January 22, 1910.

44. Victor Lougheed, Contract with John J. Montgomery to Manufacture Powered Flying Machines Based on U.S. Patent 831,173, Granted to John J. Montgomery September 18, 1906, and to Pursue Infringement of U.S. Patent 831,173, Granted to John J. Montgomery September 18, 1906, December 31, 1909 [SCU-JJM].

45. "Big Men of Finance Back of the Wrights," *New York Times*, November 19, 1909.

46. Ibid.

Chapter 15. Birds of Prey

1. Louis Blériot to the editor, *Aircraft* 1, no. 4 (1910): 66.

2. George F. Campbell-Wood, "The Wright-Curtiss-Paulhan Conflict," *Aircraft* 1, no. 4 (1910): 55.

3. "Calmly Views the Fulfillment of Life Work: Prof. J.J. Montgomery Admires Courage of Aviator, but Expresses No Surprise," *San Francisco Call,* January 24, 1910.

4. "Aeroplane Men Would Dissolve Injunction: Prof. Montgomery of Santa Clara College Assisting Curtiss," *San Jose Sunday Mercury and Herald,* January 23, 1910.

5. "Local Inventor Will Contest Claims of Famous Aeronauts," *Oakland Tribune,* January 12, 1910. See also "Montgomery Aids Fight on Wrights by a New Action," *San Francisco Examiner,* January 22, 1910.

6. Wilbur Wright to Orville Wright, September 18, 1911 (recounting his recent court testimony given in New York) [LoC-WOW].

7. Mitchell A. Wilson, *American Science and Invention: A Pictorial History* (New York: Simon & Schuster, 1954), 321.

8. "Aeroplane Father Denounces Wrights," excerpted in *New York American,* April 19, 1910.

9. Hiram Maxim to the editor, *Aircraft* 1, no. 4 (1910): 66.

10. "Octave Chanute, Who Gave Start in Aeronautics, Regrets Patent Suits," *Philadelphia Public Ledger,* January 17, 1910.

11. "Chanute Deplores Wright Patent Suits: Early Experimenter Who Turned His Data Over to Them Thinks They Will Check Progress," *New York Times,* January 23, 1910. See also "Dr. Chanute Denies Wright Flying Claim," *New York World,* January 17, 1910. With regard to "the basic warping idea," the reader will recall the second and third Montgomery gliders of 1885 and 1886, respectively.

12. Octave Chanute to Wilbur Wright, January 24, 1910, emphasis added [LoC-WOW].

13. "Patent to Aerial Association May Hit the Wrights: Principle of Side Balancing Rudders, Long in Dispute, Covered in Letters Patent Issued to Dr. Bell's Company," *New York World,* December 6, 1911.

14. Mrs. J. Herbert Sinclair, "News in General," *Aircraft* 1, no. 2 (April 1910): 71.

15. Victor Lougheed, in answer to question 21, in Court (Equity No. 33852), January 17, 1919.

16. Victor Lougheed, in answer to question 132, in Court (Equity No. 33852), January 17, 1919. "Drift" here means the total air resistance produced by the craft as measured by a spring balance as the air impacted the craft in towed flight.

17. Ibid.

18. "Montgomery Hits Wright's Patent: California College Professor Claims He Invented Warping Wings Back in 1885," *New York World,* April 24, 1910.

19. Ibid.; "Old Time Flyer Still at It," *New York Sun,* April 20, 1910; "Professor Talks Aviation," *San Francisco Examiner* (New York Bureau), April 23, 1910; John J. Montgomery, "The Origin of Warping: Professor Montgomery's Experiments," *Aeronautics* (London) 3, no. 5 (1910): 63–64; "Death of Professor John J. Montgomery," *Aeronautics* (New York) 9, no. 4 (1911): 151–54, paraphrased extensively in Charles B. Hayward, *Practi-*

cal Aeronautics: An Understandable Presentation of Interesting and Essential Facts in Aeronautical Science (Chicago: American School of Correspondence, 1912), 60–65; John J. Montgomery, "Address by Prof. John J. Montgomery: Before the Aeronautical Society, April 21, 1910," *Engineers' List* 21, no. 1 (1912): 15–21. See also "Says the Wrights Did Him Wrong," *New York Sun,* April 24, 1910.

20. B. Baden-Powell, "Warping: Professor Montgomery's Claim," Editorial, *Aeronautics* (London) 3, no. 5 (1910): 71. A transcript of Montgomery's second lecture at Chicago from 1893 had been read at the April 21, 1910, meeting of the ASNY. There is no known surviving transcript of the speech.

21. "Plan Aero Race to Go On All Day," *New York Times,* June 22, 1910.

22. "Club Notes," *Aircraft* 1, no. 6 (1910): 191.

23. "Professor Montgomery Gets Fortune for Invention," *San Jose Sunday Mercury and Herald,* March 12, 1910; Richard J. Montgomery, in answer to questions 292–294, in Court (Equity No. 33852), January 17, 1919.

24. Ibid.

25. Raymond Acre to John V. Hettich, May 6, 1946 [NASM-ELJ]. See also "California Priest Aero Expert," *Washington Post,* excerpted in the *Oakland Tribune,* June 21, 1910.

26. "Chris Buckley Loses Suit against Inventor," *San Francisco Examiner,* May 27, 1911.

27. "Aviation Notes," *Washington Post,* October 14, 1910.

28. "New Association to Be Known as the Santa Clara Valley Aero Club," *San Jose Daily Mercury,* October 23, 1910.

Chapter 16. Evergreen

1. Edmund F. Andrews, "Some Facts about Soaring Flight," *Aeronautics* (New York) 8, no. 5 (1911): 155.

2. Howard L. Scamehorn, *Balloons to Jets: A Century of Aeronautics in Illinois, 1855–1955* (Carbondale: Southern Illinois University Press, 2000), 62.

3. Ibid.

4. "Savant Working on Airship Motors," *San Francisco Call,* January 3, 1911.

5. "Professor Montgomery Tells of Electrical Inventions," *San Jose Mercury Evening News,* March 30, 1911.

6. Winsor Josselyn, notes from personal interview with Regina Montgomery at the Press Club, San Francisco, March 21, 1946 [SCU-JJM]. John's remark was relayed to Josselyn by Regina in the interview.

7. Richard J. Montgomery, in answer to questions 275–76, in Court (Equity No. 33852), January 17, 1919.

8. Victor Lougheed, *Aeroplane Designing for Amateurs* (Chicago: Reilly & Britton Co., 1912), 23.

9. "Machine Turns Turtle in Miniature Whirlwind: Accident after a Successful Flight," *San Francisco Chronicle,* November 1, 1911.

10. Hugo Gibson (Aeronautic Society of New York) to John J. Montgomery, June 29, 1911 [SCU-JJM].

11. "Plans Aeronautical Class," *Aviation,* September 1911.

12. "Personnel of Faculty of Santa Clara College Has Been Named: Prof. Montgomery, the Distinguished Scientist, and Inventor Included," *San Jose Sunday Mercury and Herald,* August 13, 1911.

13. "Club News," *Aeronautics* 8, no. 3 (1911): 110.

14. Jean-Claude Cailliez, "Le 1er vol plané international de l'histoire, par David Deluz et son Montgomery, en automne 1911" [1st International Gliding Flight in History, by David Deluz and His Montgomery in Autumn 1911], http://www.pionnair-ge.com/spip1/spip .php?article217.

15. Ibid.

16. "Wrights May Control the Air," *Daily Courier,* August 30, 1911; reprinted in the *Aberdeen Daily News,* September 18, 1911.

17. Glenn Curtiss to James E. Plew, October 20, 1911 [SCU-JJM].

18. "Aviator Walker Sues for Balance on Curtiss Motor," *San Jose Mercury Evening News,* August 23, 1912.

19. John A. Ramonda, notes, April 23, 1960, witnessed by Richard E. Neiman, April 23, 1960 [SCU-JJM].

20. Joseph C. Vierra, in answer to question 61, in Court (Equity No. 33852), January 21, 1919.

21. "Machine Turns Turtle in Miniature Whirlwind." Reinhardt is quoted extensively in the article.

22. Ibid.

23. Cornelius Reinhardt, affidavit, November 8, 1957 [SCU-JJM]. Charles D. South, a fellow professor at Santa Clara College, confirmed that John occasionally suffered from vertigo.

24. "Inquest in Case of Professor Montgomery Was Held at Santa Clara Today," *San Jose Mercury Evening News,* November 1, 1911.

25. Cornelius Reinhardt in answer to question 93, in court (Equity No. 33852), January 23, 1919.

26. Ibid.

27. Reinhardt affidavit, November 8, 1957.

28. "Glider Drops Him to Death," *Washington Post,* October 31, 1911.

29. "Machine Turns Turtle in Miniature Whirlwind."

30. Thomas S. Baldwin to Orville Wright, November 1, 1911 [SCU-JJM].

31. Orville Wright to Thomas S. Baldwin, November 18, 1911 [SCU-JJM].

32. James P. Morrisey, S.J., "Professor Montgomery: A Tribute to the Great Scientist by Rev. Father Morrisey, S.J.," *Santa Cruz Surf,* November 1, 1911.

Epilogue

1. Richard J. Montgomery, in answer to question 266, in Court (Equity No. 33852), January 25, 1919 [NASM-ELJ].

2. Alfred J. Cleary, in answer to question 82, in Court (Equity No. 33852), January 25, 1919.

3. Richard J. Montgomery to Harry F. Stabler, May 28, 1921 [NASM-JJM].

4. Alfred J. Cleary, in answer to question 5, in Court (Equity No. 33852), January 16, 1919.

5. Ibid.

6. Richard J. Montgomery, in answer to question 266, in Court (Equity No. 33852), January 25, 1919. Charles H. Lamson was a collaborator with Chanute in flying machine experiments and a designer of large human-carrying kites. Lamson held a U.S. patent for "Ribbed curved aerocurved kites" issued January 22, 1901 (No. 666,427).

7. That man was actually C. E. Warren of Tucson, Arizona.

8. Agreement between the Montgomery heirs and Frank A. Garbutt, July 27, 1914 [SCU-JJM]. Correspondence between Martin and the Montgomery heirs documenting this evolving business arrangement includes Glenn L. Martin to Richard J. Montgomery, April 27, April 28, May 10, May 24, May 25, and May 27, 1914; Richard J. Montgomery to Glenn L. Martin (telegram), May 19, 1914 [all SCU-JJM].

9. Alfred J. Cleary, in answer to question 36, in Court (Equity No. 33852), January 25, 1919.

10. Richard J. Montgomery, in answer to question 246, in Court (Equity No. 33852), January 17, 1919.

11. Alfred J. Cleary, in answer to question 5, in Court (Equity No. 33852), January 16, 1919.

12. Ibid.

13. William F. F. Durand to M. M. O'Shaughnessy, May 17, 1917 [SCU-JJM].

14. Alfred J. Cleary, in answer to question 41, in Court (Equity No. 33852), January 25, 1919.

15. Alfred J. Cleary, in answer to question 20, in Court (Equity No. 33852), January 16, 1919.

16. Alfred J. Cleary, in answer to question 30, in Court (Equity No. 33852), January 16, 1919.

17. Alfred J. Cleary, in answer to question 66, in Court (Equity No. 33852), January 25, 1919.

18. Montgomery to Stabler, May 28, 1921.

19. U.S. Government's Counsel (William D. Eakin et al.) in Equity No. 33852, "Summary of Montgomery's Work" (undated ms., ca. 1921). Orville's input is indicated in marginal notations on pp. 2 and 3 and is indicated in the text on p. 5. That the council interviewed Thomas S. Baldwin is indicated in the text (p. 21) [NASM-JJM].

20. Ibid.

21. Jerome C. Hunsaker, "Discussion of Montgomery Control" (unpublished ms., December 4, 1919) [NASM-JJM]. Spiral instability leads to a downward spinning trajectory.

22. Judge Thomas B. Booth, Chief Justice, Court of Claims of the United States, No. 33852, *Regina Cleary Montgomery, Heir, and Richard J. Montgomery, Mary C. Montgomery, Margaret H. Montgomery, and Jane E. Montgomery, Assignees of Ellen Montgomery, Heir of John J. Montgomery, Deceased, v. United States of America,* 65 Ct. Cl. 526 (May 28, 1928, decided). The comment "This is a crucial point" was made by Judge Booth [SCU-JJM].

23. Ibid.

24. Alexander Klemin, in answer to question 69, in Court (Equity No. 33852), December 20, 1919.

25. City and County of San Francisco, Board of Supervisors, *Municipal Record* 12, no. 1 (1919): 394. See also Bruce Eytinge, *Flying Guide and Log Book* (New York: John Wiley & Sons, 1921), 64, 65.

26. Godfrey L. Cabot (President, National Aeronautic Association) to Regina C. Montgomery, January 24, 1925 [SCU-JJM].

27. Gary Fogel, *Wind and Wings: The History of Soaring in San Diego* (San Diego: Rock Reef, 2000), 40.

28. Ibid., 212.

29. Reuben Fleet to Orville Wright, October 3, 1944 [SCU-JJM].

30. Orville Wright to Reuben Fleet, October 26, 1944. The header on each page includes the statement "No Part of This Personal Letter May Be Copied Without Mr. Wright's Permission" [UNC-JJM].

31. Some authors and historians in post-1944 literature have adopted Orville Wright's incorrect assertion (1944) that Chanute's published account of Montgomery's earliest flying machines of 1884–86 is *the* authentic account (Chanute, *Progress in Flying Machines* [New York: American Engineer and Railroad Journal, 1894], 248, 249). Montgomery gave the authentic primary account of his gliders and experiments at the conference at Chicago in 1893 (see chaps. 8 and 9 of this book). Chanute's discussion, stated consistently in a second-person narrative, is an interpretation of those craft/experiments. It is relevant to note that at the time, Chanute was under the understanding that Montgomery wanted to apply for patent protection on the technology. On April 21, 1910, Montgomery reprised his 1893 lecture in a speech to the Aeronautic Society of New York. He prefaced the 1910 speech by stating that Chanute had briefly described the gliders (1893, 1894), then followed with his own account of the actual experiments in the relevant context and included detail that Chanute had retained. The differences between Montgomery's primary account and Chanute's secondary account also demonstrate the natural consequences of an author serving as a secondary source on another's work: the introduction of errors and misunderstandings.

32. Albert A. Arnhym, "John J. Montgomery–The Forgotten Man of Aviation" (paper read to a meeting of the American Institute of Aeronautics and Astronautics, San Diego, Calif., November 1, 1944), San Diego Public Library, California Room, San Diego, Calif.

33. The correspondence documenting this is in the public domain and is available through the Library of Congress American Memory Collection [LoC-WOW].

34. Later released by Columbia Pictures as *Gallant Journey*.

35. Byron Morgan to Fred C. Kelly, May 24, 1946 [SCU-JJM].

36. Fred C. Kelly to Orville Wright, February 7, February 12, February 14, February 20, March 26, March 27, April 12, April 22, and May 27, 1946; Orville Wright to Fred C. Kelly, February 9, 1946 (Orville sent a three-page typed summary about Montgomery as well as a second copy of the 1944 memorandum), February 18, 1946 (a lengthy diatribe on Montgomery), March 26, 1946 (Orville reveals that Gardner and Findley are also keeping him informed on the progress of *Gallant Journey*), April 26, 1946 (a three-page letter on Montgomery), and June 1, 1946. That Orville intended his 1944 memorandum to be followed faithfully in his advocates' campaigns on the subject is evident in a letter to Earl N. Findley dated April 12, 1946, and signed by Orville's secretary, Mabel Beck. In this letter Beck communicated Orville's indignation that Findley deviated slightly from the 1944 memorandum in his article "The Montgomery Myth," *U.S. Air Services,* April 1946 [all LoC-WOW].

37. Fred C. Kelly to Harry Cohn (Columbia Pictures), March 16, 1946 [LoC-WOW].

38. Fred C. Kelly to John Hettich, May 21, 1946 [NASM-ELJ].

39. In addition to Findley, Marvin W. McFarland was a contributing editor for *U.S. Air Services.*

40. Robert L. Frey (Executive Assistant, United Press Association) to Fred C. Kelly, May 22, 1946 [LoC-WOW].

41. Edward B. Stokes to Richard J. Montgomery, December 18, 1911 [SCU-JJM]. Stokes had assisted Montgomery with flight trials adjacent to "Wheeler Hill" at Otay Mesa in 1885.

42. "The Otay Valley," *San Diego Union,* July 7, 1883.

43. Winsor Josselyn to James P. Montgomery, May 11, 1944 [SCU-JJM].

44. Frank Allen generously shared notes and correspondence that also pointed to this general area.

45. The reconstructed *Evergreen* is cataloged under NASM inventory number A19470028000. The glider is currently on display at the San Diego Air and Space Museum. The museum also holds Montgomery's handwritten original of the Soaring Flight Manuscript.

46. Fred C. Kelly, ed., *Miracle at Kitty Hawk: The Letters of Wilbur and Orville Wright* (New York: Arno, 1951).

47. Marvin W. McFarland, *The Papers of Wilbur and Orville Wright: Including the Chanute-Wright Letters and Other Papers of Octave Chanute* (New York: McGraw-Hill, 1953); C. H. Gibbs-Smith, *A History of Flying* (London: Batsford, 1953); Gibbs-Smith, *Aviation: An Historical Survey from Its Origins to the End of World War II* (London: HMSO, 1960); Fred Howard (*Wilbur and Orville: A Biography of the Wright Brothers* (New York: Ballantine, 1987). Howard also served as McFarland's technical editor for *The Papers of Wilbur and Orville Wright.*

48. This replica is currently on static display at the Hiller Aircraft Museum, San Carlos, Calif. The authors note the passing of Stan Hall on September 7, 2009.

49. Richard B. Campi, "How Scientific Were the Early Pioneers?" (unpublished ms., January 14, 1984) [SCU-JJM].

50. Ibid.

51. Bikle established a world record for sailplane altitude of 46,267 feet on February 25, 1961, over the Mojave Desert.

52. "John Montgomery: Scientist/Inventor," National Aviation Hall of Fame, http://www.nationalaviation.org/montgomery-john/. See also "John Joseph Montgomery," in James W. Jacobs, *Enshrinee Album: "The First Twenty One Years"* (Dayton, Ohio: National Aviation Hall of Fame, 1984), 134–35.

53. "Montgomery Glider Flight Competition (Reference 1883 Glider)," International Aerospace Hall of Fame.

54. Ibid.

55. Howard "Ace" Campbell, "Montgomery v. Waterman" (undated letter, ca. 1975), John J. Montgomery biographical file, San Diego Air and Space Museum Library, San Diego, Calif.

56. Mark D. Ardema and Joseph Mach, Santa Clara University School of Engineering, and William J. Adams, Jr., "John Joseph Montgomery, 1883 Glider: An International Historic Mechanical Engineering Landmark, Designated by the American Society of Mechanical Engineers, May 11, 1996, at Hiller Aircraft Museum and Santa Clara University" (brochure, 11 pp.) [SCU-JJM].

57. Norman Y. Mineta, "A Great Day In Aviation History," address in the U.S. House of Representatives, *Congressional Record,* Proceedings and Debates of the 94th Congress, First Session, Extensions of Remarks, vol. 121, April 29, 1975, 12390–91.

58. "Guidelines for Qualifications of Members of U.S. Soaring Hall of Fame," http://www.soaringmuseum.org/halloffame/procedure.html.

59. John H. McMasters, email message to Craig Harwood and Gary Fogel, April 12, 2006.

60. Preamble to meeting of the Aero Club of California, Los Angeles, October 31, 1911 [SCU-JJM].

GLOSSARY

Aileron	A movable surface for lateral control, usually found on the outer rear portion of the wing.
Airfoil	A cross-section design of a wing or lifting surface.
Airspeed	The speed of an airplane through the air.
Angle of incidence	The angle between the chord of the wing and the longitudinal axis of the fuselage on an aircraft.
Aspect ratio	The ratio of wingspan to the mean chord of the wing.
Biplane	An airplane with two wings set one above the other.
Cabane	A structure on the top of a glider fuselage used to provide additional support for the wings.
Canard	An aircraft having an elevator in front of the wing rather than behind.
Center of gravity	The center of weight of an airplane.
Chord	The width of a wing measured from the leading edge to the trailing edge.
Control surface	The surfaces that control the lateral and longitudinal movements of an airplane; the ailerons, rudder, and elevator.
Dihedral angle	The angle at which the wings rise from the fuselage relative to the wing tips.
Drag	A retarding force due to aerodynamic resistance.
Dual control	A double set of controls so that either the pilot or the passenger (typically a student) can operate the aircraft.
Elevator	A hinged, horizontal tail surface for controlling vertical movement.
Empty weight	The weight of an aircraft without pilot or equipment.
Flaps	Movable surfaces hinged to the trailing edge of the wing to facilitate approaches to landings by increasing lift and drag.
Fuselage	The main body of an aircraft that holds the pilot.
Glide	To maintain flying speed by "coasting downhill" on the air, the pull of gravity furnishing the power.
Glider	A flying machine designed for unpowered flight; the pilot uses an operable control system to guide the aircraft through the air.
Gliding angle	The angle at which the gliding descent is made.
Gross weight	The weight of an aircraft fully loaded with pilot.
Hang glider	A flying machine designed for unpowered flight; the pilot uses weight shifting to guide the aircraft through the air.
Kite	A flying tethered heavier-than-air aircraft. During the 1880s–1900s, John Montgomery referred to "kites" and "model gliders" somewhat interchangeably.
Leading edge	The foremost or entering edge of a wing.
Lift	The force exerted by the air on a plane in a direction nearly perpendicular to its motion.

Model glider   A small version of a man-carrying aircraft used for design testing, usually by hand launch and without control. In some cases, Montgomery launched model gliders from derricks to evaluate their ability to regain "equilibrium" and right themselves.

Pitch   The movement of an airplane on its lateral axis.

Rectifier   An electrical device for converting alternating current (AC) to direct current (DC).

Retrograde rotation   Rotation in the opposite direction from the typical rotation in a system.

Rheostat   A two-terminal variable resistor used to control electrical devices.

Ridge lift   The ability of a motorless aircraft to maintain or gain altitude by extracting lift generated by wind as it strikes the windward side of a hill or mountain. Also known as ridge soaring.

Roll   The movement of an airplane on its longitudinal axis.

Rudder   A vertical, hinged tail surface for directional control in a horizontal plane.

Sailplane   A streamlined flying machine capable of sustained flight without self-propulsion by extracting energy from the atmosphere in the form of lift.

Solar microscope   A microscope consisting of a mirror for reflecting sunlight through a tube, with a lens to converge the beam of light upon an object, and using another, smaller lens to magnify the object onto a screen in a dark room.

Spiral instability   The condition in which a plane tends to enter a spiral dive as a result of strong directional stability and lack of dihedral and/or yaw damping.

Stabilizer   A fixed, horizontal tail surface for vertical stability.

Tandem wing   Two wings placed one in front of the other to increase overall surface area.

Thermal soaring   The ability of a motorless aircraft to maintain or gain altitude through the extraction of lift generated by differential heating of the Earth's surface by staying within rising air currents.

Thermals   Rising air currents formed by the sun's heat radiated from the Earth.

Trailing edge   The rear edge of a wing.

Vertical fin   A fixed surface attached parallel to the longitudinal axis of an aircraft tin order to secure stability.

Wind tunnel   A research device to evaluate the effects of air moving past solid objects in the airstream, typically used to measure the lift or drag associated with such objects.

Windlass   An apparatus to move heavy weights using a winch, rope, and pulley.

Wing   The main supporting surface of the aircraft.

Wing area   The area of the supporting surfaces of an aircraft, including ailerons, but not the stabilizer or the elevator.

Wing warping   A technique to twist the wings in opposite directions to generate a change in roll.

Wingspan   The distance between the wing tips of an aircraft.

Yaw   To turn flatly from side to side.

BIBLIOGRAPHY

Archival Collections

Alameda County, Office of Clerk, Oakland, Calif.
California State Library, California Room, Sacramento, Calif.
    John Bidwell Papers
Catholic Archdiocese of San Francisco Archives, Menlo Park, Calif.
    John J. Montgomery Index
Church of Jesus Christ of Latter-day Saints, Oakland Regional Family History Center,
    Oakland, Calif.
Church of Jesus Christ of Latter-day Saints, Santa Cruz Family History Center, Santa
    Cruz, Calif.
Dayton Public Library, Local History Room, Dayton, Ohio
    Wright Brothers Collection
Gary B. Fogel, San Diego, Calif.
    Personal collection
Fordham University, Department of Archives and Special Collections, New York, N.Y.
Georgetown University Archives, Washington, D.C.
Craig S. Harwood, Santa Cruz County, Calif.
    Montgomery Family collection
Library of Congress Manuscript Division, Washington, D.C.
    Octave Chanute Papers
    Wilbur and Orville Wright Papers
Los Angeles Public Library, Main Branch, Special Collections, Los Angeles, Calif.
Oakland Public Library Main Branch, California Room, Oakland, Calif.
    John J. Montgomery biographical file
    Newspaper microform collection
San Diego Air and Space Museum Library and Archives, San Diego, Calif.
    John J. Montgomery biographical file
San Diego History Center, San Diego, Calif.
    John J. Montgomery biographical file
    Donald M. Stewart Papers
    Oral History Program (interview of Melvin D. Johnson)
San Diego Public Library, California Room, San Diego, Calif.
    John J. Montgomery biographical file
San Diego State University, South Coastal Information Center, San Diego, Calif.
    Aerial photo collection
San Francisco Public Library, Main Branch, San Francisco, Calif.
    San Francisco newspapers on microform
San Jose Public Library, Martin Luther King Branch, California Room, San Jose, Calif.
    John J. Montgomery biographical file
Santa Clara Arts and Historical Consortium, Santa Clara, Calif.
    John J. Montgomery biographical file

Santa Clara City Library, California Room, Santa Clara, Calif.
    John J. Montgomery biographical file
Santa Clara University Archives and Special Collections, University Library, Santa
    Clara, Calif.
    John J. Montgomery Collection
    Papers of Arthur Dunning Spearman, S.J., 1899–1977
Santa Cruz Public Library, Main Branch, Genealogy Room, Santa Cruz, Calif.
Save Our Heritage Organization, San Diego, Calif.
    Bruce Coons Collection
Smithsonian National Air and Space Museum, Archives Division, Washington, D.C.
    Ernest Jones Aeronautical Collection, 1906–1937
    John J. Montgomery Biographical File
Stanford University, Palo Alto, Calif.
    Cecil H. Green Library
Sutter County, Office of Clerk, Yuba City, Calif.
University of California, Berkeley, Calif.
    Bancroft Library
    Doe Memorial Library (newspaper microform collection)
University of California, Riverside, Center for Bibliographical Studies and Research,
    Riverside, Calif.
    California Digital Newspaper Collection, http://cdnc.ucr.edu/cdnc
University of California, Santa Cruz
    Special Collections, McHenry Library
    Science and Engineering Library, Map Room (historical maps)
University of Chicago, John Crerar Library, Chicago, Ill.
University of North Carolina, Chapel Hill, Louis Round Wilson Special Collections Library,
    Southern Historical Collections
    John J. Montgomery Papers, 1885–1947
    Zach Montgomery Speeches, 1864
University of North Texas Library, Denton, Texas
    Papers of President Grover Cleveland
University of Notre Dame Archives, South Bend, Ind.
    Albert Francis Zahm Papers
University of San Francisco, Archive Room, San Francisco, Calif.
U.S. Geological Survey Library, Menlo Park, Calif., Vincent E. McKelvey Building
    Historical topographic maps collection
    Historical aerial photo collection
Wings of History Air Museum, San Martin, Calif.
    John J. Montgomery biographical file
Wright State University, Special Collections and Archives, Dayton, Ohio
    Paul Laurence Dunbar Digital Collection
Yuba County Public Library, California Room, Yuba City, Calif.

Primary-Source Correspondence

Zachariah Montgomery to Richard Montgomery, August 6, 1885 [John J. Montgomery Collection, Santa Clara University Archives and Special Collections, Santa Clara, Calif., hereafter SCU-JJM]

John Montgomery to Margaret Montgomery, December 23, 1885 [SCU-JJM]

John Montgomery to James Montgomery, September 7, 1886 [SCU-JJM]

John Montgomery to Jane Montgomery, August 25, 1893 [SCU-JJM]

Octave Chanute to John Montgomery, October 24, 1893 [SCU-JJM]

John Montgomery to Octave Chanute, March 18, 1894 [there is no known surviving copy of the original; inferred from OC to JJM, March 30, 1894]

Octave Chanute to John Montgomery, March 30, 1894 [Octave Chanute Papers, Library of Congress Manuscript Division, Washington, D.C., hereafter LoC-OC]

Octave Chanute to John Montgomery, April 6, 1894 [LoC-OC]

Octave Chanute to John Montgomery, May 3, 1894 [LoC-OC]

Octave Chanute to John Montgomery, August 18, 1894 [LoC-OC]

John Montgomery to Jane Montgomery, August 30, 1894 [SCU-JJM]

Octave Chanute to John Montgomery, July 13, 1895 [LoC-OC]

Octave Chanute to John Montgomery, August 16, 1895 [LoC-OC]

Octave Chanute to John Montgomery, October 30, 1895 [LoC-OC]

Thomas S. Baldwin to John Montgomery, June 20, 1903 [SCU-JJM]

Thomas S. Baldwin to John Montgomery, September 23, 1903 [SCU-JJM]

Thomas S. Baldwin to John Montgomery, October 7, 1903 [SCU-JJM]

John Montgomery to Thomas S. Baldwin, November 15, 1903 [there is no known surviving copy of the original; inferred from OC to JJM, November 23, 1903]

Thomas S. Baldwin to John Montgomery, November 23, 1903 [SCU-JJM]

John Montgomery to Ellen Evoy Montgomery, December 15, 1903 [SCU-JJM]

Thomas S. Baldwin to John Montgomery, December 21, 1903 [SCU-JJM]

John Montgomery to James P. Montgomery, May 15, 1904 [SCU-JJM]

John Montgomery to Thomas S. Baldwin, May 24, 1904 [SCU-JJM]

John Montgomery to Fred W. Swanton, January 22, 1905 [SCU-JJM]

John Montgomery to Richard Montgomery, March 16, 1905 [SCU-JJM]

John Montgomery to Ellen Evoy Montgomery, March 20, 1905 [SCU-JJM]

Octave Chanute to John Montgomery, April 4, 1905 [SCU-JJM]

John Montgomery to Octave Chanute, April 11, 1905 [LoC-OC]

Octave Chanute to John Montgomery, April 16, 1905 [SCU-JJM]

John Montgomery to Octave Chanute, April 20, 1905 [LoC-OC]

Octave Chanute to John Montgomery, April 28, 1905 [SCU-JJM]

John Montgomery to Octave Chanute, April 30, 1905 [LoC-OC]

John Montgomery to James P. Montgomery, May 1, 1905 [SCU-JJM]

Octave Chanute to John Montgomery, May 23, 1905 [SCU-JJM]

John Montgomery to Octave Chanute, June 17, 1905 [LoC-OC]

Octave Chanute to John Montgomery, June 28, 1905 [SCU-JJM]

Octave Chanute to John Montgomery, October 28, 1905 [SCU-JJM]

John Montgomery to Octave Chanute, November 12, 1905 [LoC-OC]

Octave Chanute to John Montgomery, November 17, 1905 [SCU-JJM]

John Montgomery to Octave Chanute, November 30, 1905 [LoC-OC]

Octave Chanute to John Montgomery, December 12, 1905 [SCU-JJM]

Octave Chanute to John Montgomery, December 23, 1905 [SCU-JJM]

John Montgomery to Octave Chanute, December 29, 1905 [LoC-OC]

Octave Chanute to John Montgomery, January 10, 1906 [LoC-OC]

John Montgomery to Octave Chanute, January 29, 1906 [LoC-OC]

Octave Chanute to John Montgomery, February 16, 1906 [SCU-JJM]

Richard Montgomery to U.S. War Department, March 1906 [John J. Montgomery Biographical File, Smithsonian National Air and Space Museum, Archives Division, Washington, D.C., hereafter NASM-JJM]

U.S. War Department Board of Ordnance and Fortification to Richard Montgomery, March 20, 1906 [there is no known surviving transcript of this letter; it is discussed by Richard J. Montgomery in answer to question #126 in *Regina C. Montgomery et al. v. United States*—Equity No. 33852, January 14, 1919]

U.S. War Department to Richard Montgomery, April 13, 1906 [quoted in Richard J. Montgomery to Harry F. Stabler, May 28, 1921, NASM-JJM]

John Montgomery to James P. Montgomery, May 8, 1906 [SCU-JJM]

John Montgomery, to Octave Chanute, May 27, 1906 [LoC-OC]

Octave Chanute to John Montgomery, June 4, 1906 [SCU-JJM]

John Montgomery to Octave Chanute, July 10, 1906 [LoC-OC]

Octave Chanute to John Montgomery, December 12, 1906 [SCU-JJM]

John Montgomery to James P. Montgomery, December 16, 1906 [SCU-JJM]

Richard J. Montgomery to Harry F. Stabler, May 28, 1921 [NASM-JJM]

Books and Articles

"Accident to the Montgomery Aeroplane." *Scientific American* 93 (July 29, 1905): 82.

Aero Club of America. *Navigating the Air: A Scientific Statement of the Progress of Aëronautical Science up to the Present Time.* London: William Heinemann, 1907.

"Aeronautics." *Messenger* 44, no. 1 (1905): 102.

The Aeroplane. Santa Clara: Aeroplane Advertising Co., 1905. This book, an offprint from *The Redwood* 4, no. 8 (1905), is a compilation of articles by three different authors: Montgomery, Richard Bell, and Dennis Kavanagh. 31 pp.

Baldwin, Thomas, S. "The High Seas of Space." *National Magazine* 28, no. 4 (1908): 457–60.

Bechhold, J. H. "Die Fortschrite der Luftschiffahrt im Letzten Jahr." *Die Umschau* 10, no. 15 (1906): 284.

Bell, Richard H. "The Success of March 18." In *The Aeroplane* (Santa Clara: Aeroplane Advertising Co., 1905), 24–26.

Bill of Complaint for Infringement of Patent, *Montgomery Heirs v. Glenn H. Curtiss*, defendant. May 23, 1913. 16 pp.

Braunbeck, Gustav. *Braunbeck's Sport-lexikon: Luftschiffahrt.* Berlin: Braunbeck-Gutenberg, 1909.

Brockett, Paul. *Bibliography of Aeronautics.* Washington, D.C.: Smithsonian Institution, 1910.

Buchanan, E. L. "The Buchanan Method of Plane Making." *Aircraft* 1, no. 7 (1910): 257–58.

Burdick, John G., and Bernard J. Burdick. *Achieving Flight: The Life and Times of John J. Montgomery.* Bloomington, Ind.: Archway, 2017. Appendices contain transcriptions of several Montgomery primary source articles.

Chanute, Octave. "Montgomery." In *Pocket-Book of Aeronautics,* edited by Hermann W. L. Moedebeck, translated by W. Mansergh Varley, 309–10. London: Whittaker and Co., 1907.

———. "Progress in Flying Machines." *American Engineer and Railroad Journal* 1, no. 3 (1893): 580.

———. *Progress in Flying Machines.* New York: American Engineer and Railroad Journal Press, 1894.

Chanute, Octave, and John J. Montgomery. "Aeronautics." in *The Standard Encyclopedia: An Abridged Library and Universal Reference Book,* edited by Prof. Charles Morris. Philadelphia: Standard Encyclopedia Co., 1896.

Coupin, Henri. "Un descende de 1200 metres en aeroplane." *Le magasin pittoresque,* March 1905, 139–40.

DeMeriel, P. "Un aeroplane a 1200 meters." *La Nature,* November 25, 1905, 412.

Dienstbach, Carl. "Ein Flug von 20 Minuten mit dem Gleitapparat von Montgomery." *Illustrierte Aëronautische Mitteilungen* 9, no. 8 (1905): 255–58.

"Excitement of Success Caused Aviator's Death." *Los Angeles Times,* November 2, 1911.

"Flying-Machine Plunges into Space 4,000 Feet above the Earth: Towed Up by a Balloon and Cut-Loose—Most Daring Test Ever Made." *Popular Mechanics* 7, no. 6 (1905): 613.

"Glider Drops Him to Death." *Washington Post,* November 1, 1911.

Hargrave, Lawrence. "Aeronautics." *Journal and Proceedings of the Royal Society of New South Wales* 32 (1898): 55–65. Sydney: George Robertson & Co., 1898.

Hatch, D. S. "Aerial Navigation." In *Cyclopedia of Automobile Engineering,* vol. 4, 56–57. Chicago: American Technical Society, 1910.

Hayward, Charles B. "The Montgomery Aeroplane." *Scientific American* 92, no. 20 (1905): 404.

Hidalgo, Joseph. *History of Aerial Navigation: Lectures Delivered by Professor Joseph Hidalgo . . . before the Pacific Aero Club of San Francisco, California, January 1, 1910.* San Francisco: n.p., 1910.

Hildebrandt, Alfred. *Die luftschiffarht nach ihrer geschichtlichen und gegenwärtigen entwicklung.* Berlin: R. Oldenbourg, 1908.

Hill, Thomas A. "Status of the Wright Suit." *Aeronautics* (New York) 4, no. 10 (1909): 122–23, 164–65.

Hoernes, Hermann. *Buch des Fluges.* Vienna: G. Szelinski, 1911.

Hubbard, T. O'B., J. H. Ledeboer, and C. C. Turner. *The Aeroplane: An Elementary Text-book of the Principles of Dynamic Flight.* London: Longmans, Green and Co., 1911.

"Inquest of Professor Montgomery Was Held by Coroner at Santa Clara Today." *San Jose Mercury,* November 1, 1911.

"John J. Montgomery Falls to Death in Glider: Noted Authority Meets Lilienthal's Fate." *Fly: The National Aeronautic Monthly* 4, no. 2 (1911): 19.

Jones, Ernest L. "Death of Professor John J. Montgomery." *Aeronautics* 9, no. 4 (1911): 151–54.

Kavanagh, Dennis J. "The Flight of April 29th." In *The Aeroplane* (Santa Clara: Aeroplane Advertising Co., 1905), 27–30.

———. "The Montgomery Aeroplane: Details of the California Air Craft Which Is Attracting World-Wide Attention." *Popular Mechanics* 7, no. 7 (1905): 703–707.

"Killed by Fall in Glider: Prof. Montgomery Was Experimenting with a Motorless Monoplane." *New York Times*, November 1, 1911.

Klaussmann, A. Oskar. *Abenteuer der Luft in Ballon und Flugmaschine*. Breslau: Phönix-Verlag, 1909.

Leonard, W. H. "The Passing of a Pioneer." *Aviation* 1, no. 11 (1911): 21.

Lougheed, Victor "John J. Montgomery." *The Redwood* 9, no. 2 (1911): 60–62.

———. *Vehicles of the Air: A Popular Exposition of Modern Aeronautics with Working Drawings*. Chicago: Reilly and Britton, 1909.

Moedebeck, Hermann W. L. *Fliegende Menschen! Das Ringen um die Beherrschung der Luft mittels Flugmaschinen*. Berlin: O. Salle, 1905.

———. *Illustrierte Aëronautische Mitteilungen*. Berlin: Gustav Braunbeck & Gutenberg-Druckerei, 1907.

Montagu, Lord, ed., with a chapter by Major B. Baden-Powell. *A Short History of Balloons and Flying Machines*. London: Car Illustrated, 1907.

Montgomery, John J. "Address by Prof. John J. Montgomery: Before the Aeronautical Society, April 21, 1910." *Engineers' List* 21, no. 1 (1912): 15–21. This lecture to the Aeronautic Society of New York was published posthumously.

———. "The Aeroplane: A Scientific Study." *The Redwood* 4, no. 8 (1905): 437–68.

———. "The Aeroplane: A Scientific Study." In *The Aeroplane* (Santa Clara: Aeroplane Advertising Co., 1905), 4–15.

———. "Application of the Luminous Electrical Phenomena to Practical Illumination." Lecture delivered at St. Ignatius College, San Francisco, Calif., June 5, 1879. As described in *Saint Ignatius College Catalog for Academic Year 1878–1879* (San Francisco: Saint Ignatius College, 1879).

———. "Bases of Political Economy." *Daily Alta California*, June 4, 1878. This is a report on a lecture delivered at St. Ignatius College, San Francisco, Calif., June 3, 1878. There is no known surviving copy of the original.

———. "Discussion on the Various Papers on Soaring Flight." In *Proceedings of the Conference on Aerial Navigation, Held in Chicago, August 1, 2, 3 and 4, 1893*, edited by M. N. Forney, 247–49. Chicago: American Engineer and Railroad Journal, 1894. Reprise of the first of two lectures delivered by Montgomery to the International Conference on Aerial Navigation, Chicago, August 1, 1893.

———. "Discussion on the Various Papers on Soaring Flight." *Aeronautics* (New York) 1, no. 10 (1894): 127–28.

———. "The Mechanical Principles of a Bird's Wing Involved in Soaring and Their Relation to Aeronautics." *Los Angeles Herald*, November 10, 1897. Lecture delivered at the Southern California Academy of Science, Los Angeles, December 6, 1897. There is no known surviving copy of the original.

———. "New Principles in Aerial Flight." *Scientific American Supplement* 1560 (1905): 24991–93.

———. "The Origin of Warping: Professor Montgomery's Experiments." *Aeronautics* (London) 3, no. 6 (1910): 63–64. This is an excerpt from a speech delivered to the New York Aeronautical Society, New York, April 21, 1910. It was also excerpted in "Death of Professor John J. Montgomery," *Aeronautics* (New York) 9, no. 4 (1911): 151–54, and was printed posthumously in full in the *Engineers' List* 21, no. 1 (1912): 15–21. It is also paraphrased extensively in Charles B. Hayward, *Practical Aeronautics: An Understandable Presentation of Interesting and Essential Facts in Aeronautical Science* (Chicago: American School of Correspondence, 1912), 61–64.

———. "The Origin of Warping." Lecture delivered to the Aeronautic Society of New York, April 21, 1910. This was a reprise of a lecture delivered to the International Conference on Aerial Navigation, Chicago, August 1, 1893. It was excerpted in several publications between 1910 and 1912. There is no known surviving copy of the original lecture text.

———. "Our Tutors in the Art of Flying." *Aeronautics* (New York) 8, no. 9 (1915): 99–100. Published posthumously.

———. "Principles Involved in the Formation of Wing Surfaces and the Phenomenon of Soaring." Lecture delivered to the International Aeronautics Congress, New York, October 29, 1907. There is no known surviving copy of the original lecture text, but it was published in *Aeronautics* (see next entry).

———. "Principles Involved in the Formation of Wing Surfaces and the Phenomenon of Soaring." *Aeronautics* (New York) 3, no. 4 (1908): 30–33; 3, no. 5 (1908): 39–40; 3, no. 6 (1908): 32–36; 4, no. 1 (1909): 43–46.

Short, Simine. *Locomotive to Aeromotive: Octave Chanute and the Transportation Revolution.* Urbana: University of Illinois Press, 2011.

———. "Some Early Gliding Experiments in America." *Aeronautics* (New York) 4, no. 1 (1909): 47–50.

"Montgomery Hits Wright's Patent: California College Professor Claims He Invented Warping Wings Back in 1885." *New York World,* April 24, 1910.

Nimführ, Raimund. *Leitfaden der Luftschiffahrt und Flugtechnik.* Leipzig: A. Hartleben, 1910.

"Notizen." *Wiener Luftschiffer-zeitung* 4, no. 8 (1905): 169–70.

Nunan, Thomas, J. "A Flying Machine That Flies." *Motor Magazine* 4, no. 3 (1905): 48–49.

Peyrey, François. *Les oiseaux artificiels.* Paris: H. Dunod et E. Pinat, 1909.

"Pioneer Inventor Is a Martyr." *Aero Magazine* 3, no. 6 (1911): 116.

"Plans Aeronautical Class." *Aviation,* September 1911.

"Plunge Kills Montgomery." *San Francisco Call,* November 1, 1911.

"Professor Montgomery's Experiments." *Aeronautics* (London) 3, no. 6 (1910): 73, 111.

"Recent Experiments with Aeroplanes" *Aeronautical Journal,* July 1905, 52.

Root, Amos I. "Flying Machines Up to Date: Wright Brothers Still Ahead." *Gleanings in Bee Culture,* April 1, 1905.

Soreau, Rodolphe. *État actuel et avenir de l'aviation.* Paris: F. L. Vivien, 1909.

Tandy, Edward T. *An Epitome of the Work of the Aeronautic Society of from July, 1908, to December, 1909.* New York: Aeronautic Society of New York, 1910.

"Test of Aeroplane Is Death to Experimentor." *Los Angeles Times,* November 1, 1911.

Turner, Charles C. *Aerial Navigation of Today: A Popular Account of the Evolution of Aeronautics.* London: Seeley & Co., 1910.

Veuve, William P. "John J. Montgomery." *The Redwood* 9, no. 2 (1911): 44–59.

Walker, Thomas, T. O'B. Hubbard, and J. H. Ledeboer. *The Art of Flying.* [London]: Printed and published for the Aeronautical Society of Great Britain by King, Sell & Olding.

Wheeler, Edward J. "An Aeroplane That Actually Soars." *Current Literature* 39 (1905): 184–86.

Zahm, Albert F. *Aërial Navigation: A Popular Treatise on the Growth of Air Craft and on Aëronautical Meteorology.* New York and London: D. Appleton, 1911.

———. "Octave Chanute: His Work and Influence." *Scientific American* 104 (1911): 463.

———. "The Undeveloped Art of Soaring: Its Theory and Practice." *Scientific American Supplement* 1872 (November 18, 1911): 332–33.

Select Publications and Lectures That Mention John Joseph Montgomery (in Chronological Order, 1912–Present)

Lougheed, Victor. *Aeroplane Designing for Amateurs.* Chicago: Reilly and Britton, 1912.

Hayward, Charles B. *Practical Aeronautics: An Understandable Presentation of Interesting and Essential Facts in Aeronautical Science.* Chicago: American School of Correspondence, 1912.

Verrill, A. Hyatt. *Harper's Aircraft Book: Why Aeroplanes Fly, How to Make Models, and All about Aircraft, Little and Big.* New York: Harper & Brothers. 1913.

South, Charles D., Sr. "The Father of the Aeroplane." *The Redwood* 14, no. 1 (1914): 2–11.

"John Joseph Montgomery." In *The National Cyclopaedia of American Biography,* edited by George Derby et al. New York: James T. White Co., 1916.

"An American Pioneer in of Soaring Flight: John J. Mongtomery." *Aviation* 4, no. 1 (1918): 302–303.

Redman, Leander A. *Professor Montgomery's Discoveries in Celestial Mechanics.* San Francisco: Pernau-Walsh Printing Co., 1919.

Hunsaker, Jerome C. "Discussion of Montgomery Control." Unpublished manuscript, 1919. John J. Montgomery Biographical File, Smithsonian National Air and Space Museum, Archives Division, Washington, D.C.

Colwell, J. H. "The Origin and Development of Aeronautics." *Journal of the Patent Office Society* 3 (1920): 12.

Jones, Ernest L. "Chronology of Aviation." Unpublished manuscript, compiled from approximately 1921 and continued over an unknown period of years prior to 1946. John J. Montgomery Biographical File (Folders 1 and 2), Ernest Jones Aeronautical Collection, Smithsonian National Air and Space Museum, Archives Division, Washington, D.C.

Eytinge, Bruce. *Flying Guide and Log Book.* New York: John Wiley & Sons, 1921.

Eakin, William. "Summary of Montgomery's Work." Unpublished manuscript, ca. 1921. *Montgomery Heirs v. United States,* U.S. Court of Claims, No. 33852. John J. Montgomery Biographical File, Smithsonian National Air and Space Museum, Archives Division, Washington, D.C. This document was generated by the U.S. government's defense team for guidance in their closing arguments in the court case.

McGuire, C. E., ed. *Catholic Builders of the Nation: A Symposium on the Catholic Contribution to the Civilization of the United States.* Boston: Continental Press, 1923.

Kende, George. "Three Pioneers in Aviation: S. P. Langley, Otto Lilienthal, and John J. Montgomery." Bachelor's thesis, University of California, Berkeley, 1928. Doe Library, University of California, Berkeley.

Davy, M. J. B. *Handbook of the Collections Illustrating Aeronautics*. Vol. 1: *Heavier-Than-Air Craft: A Brief Outline of the History and Development of Mechanical Flight with Reference to the National Aeronautical Collection, and a Catalogue of the Exhibits*. London: Science Museum, 1929.

Fogel, Gary B., and Craig S. Harwood. "John J. Montgomery's Circulation Theory of Lift." 54th Aerospace Sciences Meeting (SciTech, 2016). American Institute of Aeronautics and Astronautics, San Diego, Calif., January 4–8, 2016. Paper no. AIAA 2016-1599.

Harwood, Craig S., and Gary B. Fogel. "On the Invention of Lateral Control." Aerospace Sciences Meeting (SciTech, 2016). American Institute of Aeronautics and Astronautics, San Diego, Calif., January 7–11, 2019. Paper no. AIAA 2019-0123.

Martin, Robert E. "American Unknown to Fame, First Man on Record to Leave Earth on Wings." *Popular Science Monthly* 117, no. 4 (1930): 19, 21, 145–47.

Miller, Francis T. *The World in the Air: The Story of Flying in Pictures*. Vol. 2. New York: G. P. Putnam, 1930.

Teale, Edwin Way. *The Book of Gliders*. New York: E. P. Dutton, 1930.

Hunt, Rockwell D. "John J. Montgomery." *California and Californians* 3 (1932): 126–27.

Derleth, Charles, Jr. "The History of Flight." Lecture delivered to the Montgomery Day Convocation, University of Santa Clara, March 18, 1934. John J. Montgomery Collection, Santa Clara University Archives and Special Collections.

———. "The History of Flight." *The Owl*, May 1934, 8–9, 15.

Davy, M. J. B. *Aeronautics*. London: Science Museum, 1935.

Miller, H. B. "A Forgotten Pioneer." *Model Airplane News*, June 1936, 6–7, 42–44.

Jones, Ernest L. "Montgomery and His Nineteenth Century Glider." *CHIRP* 20 (June 1937), http://www.earlyaviators.com/emontgom.htm.

Davy, M. J. B. *Interpretive History of Flight*. London: HMSO, 1937.

Dollfus, Charles, and Henri Bouché. *Histoire de l'aéronautique*. Paris: L'Illustration, 1938.

Josselyn, Winsor. "He Flew in 1883." *Harper's Magazine*, June 1940, 28–31.

Crowder, Farnsworth. "California's Forgotten Eagle." *Westways* 33, no. 12 (1941): 12–13, 17–19.

Hirth, Wolf. *Manual del vuelo a vela*. Barcelona: Editorial Labor, 1942.

Milbank, Jeremiah, Jr. *The First Century of Flight in America: An Introductory Survey*. Princeton: Princeton University Press, 1943.

Klemin, Alexander. "Gliding." In *Encyclopedia Britannica*, vol. 10, 428–29. Chicago: Encyclopedia Britannica Co., 1943.

Stieri, Emanuele. *Gliders and Glider Training*. New York: Duell, Sloan and Pearce, 1943.

Keen, Harold. "Father of Flight." *Skyways*, August 1944, 40.

Arnhym, Albert A. "John J. Montgomery—The Forgotten Man of Aviation." Paper read to a meeting of the American Institute of Aeronatics and Astronautics, San Diego, Calif., November 1, 1944. San Diego Public Library, California Room, San Diego, Calif.

Marshall, Jim. "Wing Talk." *Colliers* 115 (1945): 8–10.

Hettich, John. "Montgomery and the Birds." *Aero Digest* 52, no. 1 (1946): 40, 124.

"Forgotten Pioneer." *Flying* 39, no. 3 (1946): 52–53, 82, 84.

Wolfe, Ruth. *Pioneering in Aviation*. New York: Charles E. Merrill Co., 1946.

Burlingame, Roger. *Inventors behind the Inventor*. New York: Harcourt, Brace, and World, 1947.

Freudenthal, Elsbeth Estelle. *Flight into History: The Wright Brothers and the Air Age.* Norman: University of Oklahoma Press, 1949.

Kelly, Fred C. "Myths of Aviation: Early Gliding Experiments of John J. Montgomery Are Evaluated against a Background of Seven Decades of Progress in Aeronautics." *Technology Review* 52 (December 1949): 92, 110.

———. "Montgomery's Great Glide, Aviation's Myth." *Science Digest* 27, no. 4 (1950): 81–83.

Rhodes, W. T., and A. K. Gilbert. "Montgomery Freeway." *California Highways,* January–February 1951, 34–35.

Smith, Myles W. "A Man Alone: The American Who Flew before the Wright Brothers." *Air Age* 6, no. 7 (1951): 6–8.

Kelly, Fred C. *Miracle at Kitty Hawk: The Letters of Wilbur and Orville Wright.* New York: Farrar, Straus and Young, 1951.

McFarland, Marvin W., ed. *The Papers of Wilbur and Orville Wright: Including the Chanute-Wright Letters and Other Papers of Octave Chanute.* New York: McGraw-Hill, 1953.

Gibbs-Smith, Charles H. *A History of Flying.* London: B. T. Batsford, 1953.

Historical Data: A Chronology of American Aviation Events, 1903–1953. Washington, D.C.: Department of the Air Force, 1953.

Buchanan, Lamont. *The Flying Years: A Pictorial History of Man's Conquest of the Air.* New York: Putnam & Sons, 1953.

Hall, Courtney Robert. *History of American Industrial Science.* New York: Library Publishers, 1953.

Woytinsky, W. S., and E. S. Woytinsky. *World Commerce and Governments: Trends and Outlook.* New York: Twentieth Century Fund, 1955.

Frederick, John H. *Commercial Air Transportation.* Homewood, Ill.: Richard D. Irwin, 1955.

Garber, Paul E. *The National Aeronautical Collections, Smithsonian Institution, National Air Museum.* 9th ed. Washington, D.C.: National Air and Space Museum, 1956.

Reinhardt, Cornelius. Letter titled "Witness of John Joseph Montgomery Flights and of His Tragic Death." November 8, 1957. Cornelius Reinhardt Papers, John J. Montgomery Collection, Santa Clara University Archives and Special Collections, Santa Clara, Calif.

Scamehorn, Howard L. *Balloons to Jets: A Century of Aeronautics in Illinois, 1855–1955.* Carbondale: Southern Illinois University Press, 1957.

Mossman, Frank H., and Newton Morton. *Principles of Transportation.* New York: Ronald Press Co., 1957.

Hennessy, Juliette A. *The United States Army Air Arm, April 1861 to April 1917.* Maxwell Air Force Base, Ala.: Office of Air Force History, United States Air Force, 1958.

Ward, Baldwin H. *Flight: A Pictorial History of Aviation.* Los Angeles: Year Magazine, 1958.

Cole, Martin. "Pioneer Birdman." *California Herald* 6, no. 5 (1959): 3.

Thompson, Jack. "Relating to the 50th Anniversary of the Death of John J. Montgomery." Senate Resolution No. 37, *Senate Journal,* March 15, 1960, 213–14.

Gibbs-Smith, Charles. H. *The Aeroplane: An Historical Survey of Its Origins and Development.* London: HMSO, 1960.

Hall, Stanley A. "Montgomery Project Plan." Working paper, Lockheed Aircraft Corporation, April 17, 1961. John J. Montgomery Collection, Santa Clara University Archives and Special Collections, Santa Clara, Calif.

White, Lynn, Jr. "Eilmer of Malmesbury, an Eleventh Century Aviator: A Case Study of Technological Innovation, Its Context and Tradition." *Technology and Culture 2*, no. 2 (1961): 97–111.

Rae, John B. "Science and Engineering in the History of Aviation." *Technology and Culture 2*, no. 4 (1961): 391.

Campi, Richard B. "Description and Analysis of the 1911 Montgomery Controllable Man Carrying Glider." Working paper, December 29, 1961. Richard Campi Papers, SCU-JJM.

Cole, Martin. "The Last Flight of John Joseph Montgomery." *American Aviation Historical Society Journal,* Winter 1961, 282–84.

Lambie, Jack L. "John J. Montgomery." Unpublished manuscript, 1961. SCU-JJM.

Johnson, Kenneth M. *Aerial California: An Account of Early Flight in Northern and Southern California, 1849 to World War I.* Los Angeles: Dawson's Book Shop, 1961.

Wilkie, David F. "He Knew Montgomery." *The Current News* 11, no. 4 (1962): 8–9.

Hawes, William. "The Pioneer Development of Aviation in San Diego." Master's thesis, San Diego State College, 1962.

Zacher, Hans. "The Shape of High Performance Sailplane Technical Development." In *The World's Sailplanes: Die Segelflugzeuge der Welt. Les planeurs du monde,* vol. 3, edited by B. S. Shenstone and K. G. Wilkinson, 99. Zurich: Organisation scientifique at et technique internationale du vol à voile (OSTIV) and Schweizer Aero-Revue/Aero-Revue Suisse, 1963.

Cole, Martin. "Last Flight." *Sport Aviation,* March 1964, 31–32.

Means, James Howard. *James Means and the Problem of Man-flight, during the Period 1882–1920.* Washington, D.C.: Smithsonian Institution, 1964.

"In Memoriam: Cleve T. Shaffer." *CHIRP,* no. 17 (December 1964). http://earlyaviators.com/eshaffer.htm.

Garber, Paul E. *The National Aeronautical Collections, Smithsonian Institution National Air Museum.* 10th ed. Washington, D.C.: Smithsonian Press, 1965.

Welch, Ann, and Lorne Welch. *The Story of Gliding.* London: Murray, 1965.

Dollfus, Charles, Henry Beaubois, and Camille Rougeron, eds. *L'homme, l'air, et l'espace (aeronautique-astronautique).* Paris: Éditions de l'illustration, 1965.

Emme, Eugene M. *A History of Space Flight.* New York: Holt, Rinehart and Winston, 1965.

Cole, Martin. "The Last Flight of John Joseph Montgomery." *A.A.H.S. Journal,* Winter 1967, 282–84.

Morgan, Jane. *Electronics in the West: The First Fifty Years.* Palo Alto: National Press Books, 1967.

Spearman, Arthur Dunning, S.J. *John Joseph Montgomery, 1858–1911: Father of Basic Flying.* Santa Clara: University of Santa Clara, 1967. Revised ed., 1977.

———. "The Story behind the Book: John Joseph Montgomery, 1858–1911—Father of Basic Flying." *Soaring* 32, no. 3 (1968): 10–12.

Jackson, Eugene E. "The Shape of His Wings." *AOPA Pilot,* December 1968, 68–72.

Dwiggins, Don. *On Silent Wings: Adventures in Motorless Flight.* New York: Grosset & Dunlap, 1970.

Rogallo, Francis M., Delwin R. Croom, and William C. Sleeman, Jr. "Flexible Wings for

Transportation." Paper presented at International Congress of Transportation Conferences, Society of Automotive Engineers, Washington, D.C., May 31–June 2, 1972. *S.A.E. Transactions* 81, pt. 3 (1973): 1781.

Verbarg, Leonard H. *Celebrities at Your Doorstep: A Selection of Knave Personalities from the Sunday Tribune.* Berkeley: Alameda County Historical Society, 1972.

Connick, George P. "John J. Montgomery, His Life in Brief." In California Pioneers of Santa Clara County, *Santa Clara County Pioneer Papers, 1973,* 30–55. San Jose: Smith & McKay, 1973.

Hanson, Norwood Russell. *What I Do Not Believe: And Other Essays.* Edited by Stephen Toulmin and Harry Woolf. Dordrecht: D. Reidel, 1974.

Wragg, David W. *Flight before Flying.* Berkshire: Osprey Publishing, 1974.

Campi, Richard B. "John J. Montgomery: Details of Engineering Excellence." Address delivered to a meeting of the Experimental Aircraft Association, Sunnyvale, Calif., April 16, 1975. SCU-JJM.

Mineta, Norman I. "A Great Day In Aviation History." In *Congressional Record,* Proceedings and Debates of the 94th Congress, First Session, Extensions of Remarks, vol. 121, April 29, 1975, 12390–91.

———. "John Joseph Montgomery: A Chronology of Achievements." March 1976. SCU-JJM.

Mrazek, James E. *Hang Gliding and Soaring: A Complete Introduction to the Newest Way to Fly.* New York: St. Martin's Press, 1976.

Haining, Peter. *The Compleat Birdman: An Illustrated History of Man-Powered Flight.* London: Robert Hale, 1976.

Andrews, Allen. *Back to the Drawing Board: The Evolution of Flying Machines.* Newton Abbot: David and Charles, 1977.

Emme, Eugene M. *Two Hundred Years of Flight in America: A Bicentennial Survey.* San Diego: American Astronautical Society, 1977.

Williams, George E. "Reach for the Restless Wind." Unpublished manuscript, 1979 (revised 1986). SCU-JJM.

Weston, George. "John Joseph Montgomery." *Aviation Quarterly* 5, no. 3 (1979): 247–67.

Supf, Peter, and Georg Brütting. *Das Buch der deutschen Fluggeschichte.* Stuttgart: Drei Brunnen Verlag, 1979.

Otto Lilienthal and Octave Chanute: Pioneers of Gliding. Washington, D.C.: Smithsonian Institution, National Air and Space Museum, 1980.

Weston, George. "He Made the First Controlled Winged Flight: Western Aviator John Montgomery." *American West,* July/August 1981, 48–54, 66–67.

Crouch, Tom D. *A Dream of Wings: Americans and the Airplane, 1875–1905.* Washington, D.C.: Smithsonian Institution Press, 1981.

Chen, M. K., and J. H. McMasters. "From Paleoaeronautics to Altostratus: A Technical History of Soaring." Paper presented at Aircraft Systems and Technology Conference, American Institute of Aeronautics and Astronautics, Dayton, Ohio, August 11–13, 1981. Paper No. AIAA-1981-1611.

Chana, William F. "Kitty Hawk to 2003." Paper presented at Aircraft Systems and Technology Conference, American Institute of Aeronautics and Astronautics, Dayton, Ohio, August 11–13, 1981. Paper No. AIAA-1981-1615.

Williams, George E. "The Pink Maiden." *Soaring,* April 1982, 17–20.

Ryan, Bertha M. "An Overview of the Development of Soaring." *Soaring,* May 1982, 62–64.

Slasinski, Carolyn. "Gliders and the Law." *American Journal of Comparative Law Supplement* 30 (1982): 329.

"1883–1983: First Aviation Centennial." *Chula Vista Historical Society Bulletin* 1, no. 2 (1983): 1–12.

McGinty, Brian. "Wings over Otay." *American History* 19, no. 8 (1984): 30–33.

Schoneberger, William A., and Paul Sonnenburg. *California Wings: A History of Aviation in the Golden State.* Woodland Hills, Calif.: Windsor Publications, 1984.

Jacobs, James W. "John Joseph Montgomery." In James W. Jacobs, *Enshrinee Album: "The First Twenty-One Years,"* 134–35. Dayton, Ohio: National Aviation Hall of Fame, 1984.

Larsen, O. H. "Montgomery Glider—1883." Southern California Historical Aviation Foundation, April 30, 1985.

Boyne, Walter J. *The Smithsonian Book of Flight.* Washington, D.C.: Smithsonian Books, 1987.

Howard, Fred. *Wilbur and Orville: A Biography of the Wright Brothers.* New York: Ballantine Books, 1987.

Crouch, Tom D. *The Bishop's Boys: A Life of Wilbur and Orville Wright.* New York: W. W. Norton & Co., 1989.

Kelley, Herbert L. "Montgomery." *WW1 Aero: The Journal of the Early Aeroplane* 124 (May 1989): 44–59.

———. "Montgomery." *WW1 Aero: The Journal of the Early Aeroplane* 131 (February 1991): 39–49.

Wegener, Peter, P. *What Makes Airplanes Fly? History, Science, and Applications of Aerodynamics.* New York: Springer-Verlag, 1991.

"John Joseph Montgomery, 1883 Glider: An International Historic Mechanical Engineering Landmark Designated by the American Society of Mechanical Engineers, May 11, 1996, at Hiller Aircraft Museum and Santa Clara University." Brochure, 11 pp. http://www.asme.org/about-asme/history/landmarks/topics-a-l/air-and-space-transportation/-189-montgomery-glider-(1883).

Yenne, Bill. *Legends of Flight: With the National Aviation Hall of Fame.* Lincolnwood, Ill.: Publications International, 1997.

McCormick, Barnes W. "The Evolution of Technology and Education in Applied Aerodynamics." 37th Aerospace Sciences Meeting and Exhibit, American Institute of Aeronautics and Astronautics, Reno, Nev., January 11–14, 1999. Paper No. 99–0117.

Teacher's Guide for Aerospace: The Journey for Flight. Maxwell Air Force Base, Ala.: Civil Air Patrol National Headquarters, 2000.

Thurston, David B. *The World's Most Significant and Magnificent Aircraft: Evolution of the Modern Airplane.* Warrendale, Pa.: Society of Automotive Engineers, 2000.

Fogel, Gary. *Wind and Wings: The History of Soaring in San Diego.* San Diego: Rock Reef, 2000.

Weber, Francis J. *Encyclopedia of California's Catholic Heritage, 1769–1999.* Mission Hills, Calif.: St. Francis Historical Society, 2000.

McCormick, Barnes W. "The Growth of Aerospace Education Following Its Beginning." 40th AIAA Aerospace Sciences Meeting and Exhibit, American Institute of Aeronautics and Astronautics, Reno, Nev., January 14–17, 2002. Paper No. 2002–560.

Dees, Paul. "The Rebirth of Hang Gliding and Ultralight Sport Aviation." AIAA International Air and Space Symposium and Exposition: The Next 100 Years, American Institute of Aeronautics and Astronautics, Dayton, Ohio, July 14–17, 2003. Paper No. 2003–2876.

Fogel, Gary B., and Craig S. Harwood. "John J. Montgomery and The First Gliding Flights in America, 1884–1886." *WWI Aero* 202 (2010): 21–26.

Flandro, Gary, A., "Some Forgotten Aeronautical History." Speech to the Arnold Association of Professional Societies, AAPS/AIAA Luncheon, Arnold Air Force Base, Tenn., February 23, 2011.

Patents Held by John J. Montgomery

1884. Devulcanizing and restoring vulcanized rubber. U.S. Patent 308,189. Filed May 27, 1884, issued November 18, 1884.

1895. Petroleum burner. U.S. Patent 549,679. Filed June 25, 1895, issued November 12, 1895.

1895. Petroleum burner and furnace. British Patent 21,477. Issued November 12, 1895.

1895. Petroleum oven. German Patent 88,977. Issued November 12, 1895.

1895. Petroleum burner. Canadian Patent 50,585. Issued November 14, 1895.

1901. Concentrator. Canadian Patent 70,319. Issued February 19, 1901.

1900. Concentrator. U.S. Patent 679,155. Filed June 6, 1900, issued July 23, 1901.

1902. Concentrator. U.S. Patent 742,889. Filed June 13, 1902, issued November 3, 1903.

1905. Aeroplane. U.S. Patent 831,173. Filed April 26, 1905, issued September 18, 1906.

1909. Rectifying electric currents. U.S. Patent 974,171. Filed August 28, 1909, issued November 1, 1910.

1909. Process for compelling electric motors to keep in step with the waves or impulses of the current driving them, and a motor embodying the process. U.S. Patent 974,415. Filed August 28, 1909, issued November 1, 1910.

INDEX

References to illustrations appear in italics.

CPSIA information can be obtained
at www.ICGtesting.com
Printed in the USA
LVHW042309040320
648991LV00002B/246